Lecture Notes in Electrical Engineering

Volume 277

About this Series

"Lecture Notes in Electrical Engineering (LNEE)" is a book series which reports the latest research and developments in Electrical Engineering, namely:

- Communication, Networks, and Information Theory
- Computer Engineering
- Signal, Image, Speech and Information Processing
- Circuits and Systems
- Bioengineering

LNEE publishes authored monographs and contributed volumes which present cutting edge research information as well as new perspectives on classical fields, while maintaining Springer's high standards of academic excellence. Also considered for publication are lecture materials, proceedings, and other related materials of exceptionally high quality and interest. The subject matter should be original and timely, reporting the latest research and developments in all areas of electrical engineering.

The audience for the books in LNEE consists of advanced level students, researchers, and industry professionals working at the forefront of their fields. Much like Springer's other Lecture Notes series, LNEE will be distributed through Springer's print and electronic publishing channels.

More information about this series at http://www.springer.com/series/7818

Hiroyuki Kajimoto · Hideyuki Ando
Ki-Uk Kyung
Editors

Haptic Interaction

Perception, Devices and Applications

Springer

Editors
Hiroyuki Kajimoto
The University of Electro-Communications
Tokyo
Japan

Hideyuki Ando
Osaka University
Osaka
Japan

Ki-Uk Kyung
Electronics and Telecommunications Research Institute
Seoul
Korea, Republic of (South Korea)

ISSN 1876-1100 ISSN 1876-1119 (electronic)
Lecture Notes in Electrical Engineering
ISBN 978-4-431-55689-3 ISBN 978-4-431-55690-9 (eBook)
DOI 10.1007/978-4-431-55690-9

Library of Congress Control Number: 2015938316

Springer Tokyo Heidelberg New York Dordrecht London

Printed on acid-free paper

Springer Japan KK is part of Springer Science+Business Media (www.springer.com)

Preface

We are pleased to present the proceedings of the first international conference, Asia Haptics 2014, held in Tsukuba, Japan, November 18–20.

Asia Haptics is a new type of international conference for the haptics fields, featuring interactive presentations with demonstrations. We received 62 submissions, and the review process led to 58 of those being accepted for publication. With the addition of 10 late-breaking demonstrations and 3 exhibits, the conference became the place to experience 71 live demonstrations of haptics research.

While the haptics-related research field is huge, the book divides it into six parts.

Part I is composed of nine chapters, treating perception and psychophysics of haptics. They are unquestionably the basis of haptics research.

Part II comprises 12 chapters, dealing with tactile devices for skin sensation, such as vibration, pressure, and temperature, and their rendering methods.

Part III consists of 13 chapters, considering force feedback devices and rendering methods. Some of them utilized tactile-based illusions, so the distinction between tactile devices and force feedback devices is vague.

Part IV is composed of five chapters, concerned with sensors such as the pressure distribution sensor and force sensor, many of which enable novel computer–human interaction.

Part V is made up of seven chapters, going into medical applications including surgery simulation and rehabilitation.

Part VI contains 11 chapters, discussing the application of haptics to virtual reality, telepresence, and multimedia, all exploring new application areas of haptics.

This book helps not only active haptic researchers but also general readers to understand what is going on in this interdisciplinary area of science and technology. All chapters have accompanying videos available online (Springer video). Therefore, readers can easily understand the concept of the work with the supplemental videos.

December 2014

Hiroyuki Kajimoto
Hideyuki Ando
Ki-Uk Kyung

Contents

Part I
Perception

The Scaling of the Haptic Perception on the Fingertip Using an Interface of Anthropomorphic Finger Motions

Yusuke Ujitoko and Koichi Hirota

Abstract The demonstration described in this paper attempts to give users tactile feedback on the sole of the virtual avatar using the locomotion interface of anthropomorphic finger motions. We believe that the illusion in the contact area can be caused by the sense of ownership derived from the close relationship of the motion between the fingers and the avatar's legs. The objective of this study was to prove the possibility that fingers can be substituted in the place of legs in locomotion interfaces in terms of tactile sensation.

Keywords Finger motion · Locomotion interface · Cross-modal · Tactile sensation

1 Introduction

Studies on structuring the natural general-purpose locomotion interface that reflect the walking and running motion of the avatar in a virtual environment (VE) have been conducted. One method for operating a virtual avatar is to use fingers instead of legs. For example, finger motions that pantomime leg movements can be used as inputs for operating bipedal walking robots [1] and for navigation [2]. However, to the best of our knowledge, no study has been conducted that focuses on tactile sensations involved when a user controls a virtual avatar. An essential factor of locomotion using legs is the sense of contact between the feet and the ground

Y. Ujitoko (✉)
Graduate School of Interdisciplinary Information Studies, The University of Tokyo, 7-3-1 Hongo, Bunkyo-ku, Tokyo, Japan
e-mail: yusuke.ujitoko@gmail.com

K. Hirota
Interfaculty Initiative in Information Studies, The University of Tokyo, 7-3-1 Hongo, Bunkyo-ku, Tokyo, Japan
e-mail: k-hirota@k.u-tokyo.ac.jp

H. Kajimoto et al. (eds.), *Haptic Interaction*, Lecture Notes in Electrical Engineering 277, DOI 10.1007/978-4-431-55690-9_1

surface. Our motivation is to present this sense of contact through the contact between the fingers used for making anthropomorphic finger motions and the touch interface.

We believe that the illusory ownership of the virtual avatar can be induced when active finger motions perceived by the somatic senses synchronize the avatar's leg movement displayed to the users. Under this condition, users are expected to interpret their fingers as avatar's legs. Furthermore, interpretation on the tactile stimulation on the fingers as the feeling of touch on the feet can be also expected.

Our hypothesis predicts that if a user is absolutely convinced that the tactile sensation experienced by fingers is the same as that of the legs, the scale of the tactile sensation of the texture can be extended to one that is equivalent to the sole of the foot. For example, when fingers move on gravel, users feel that they are moving their legs on a large rock. If this relationship is clarified, we can decide the texture to be displayed for the fingers based on the ground texture in the VEs.

2 Demonstration

In this demonstration, participants "walk" by using anthropomorphic finger motions across artificial tactile boards with individual raised portions of various diameters (Fig. 1). A participant's palm and fingers are tracked by the Leap Motion Controller. The Unity3D game engine is used to obtain these data and display the first-person view. Two types of feedback are provided by the system: the graphical perspective rendered from the viewpoint of the avatar and the sense of touch from the tactile board. In order to test the hypothesis, we investigate the effect of an

Fig. 1 Artificial tactile board

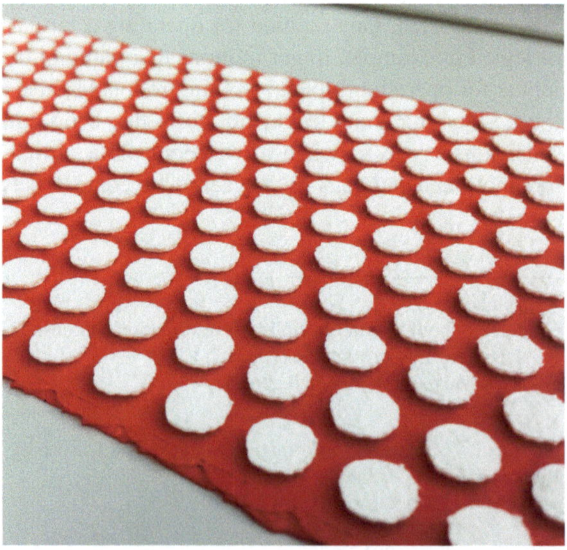

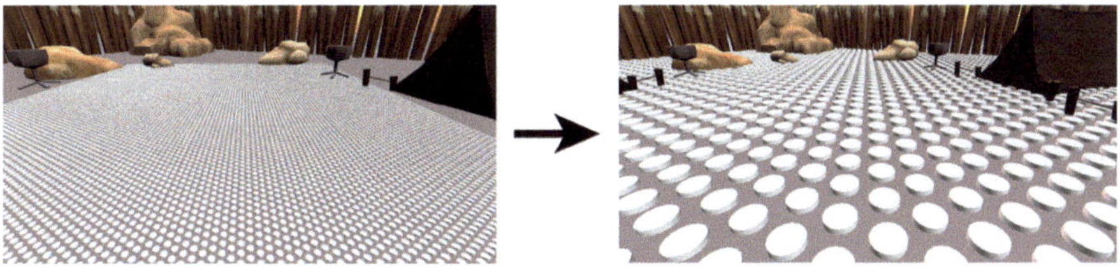

Fig. 2 Participants operate the virtual ground texture

avatar's sole size. However, in this demonstration, the variant is not the avatar's sole size, but the avatar's body size. This is because, in general, we are not conscious of sole size. It is theoretically expected that the enlargement ratio of the tactile sensation would be proportional to the body size of the avatar. We presume that it happens occasionally and depends on the functioning of the immersion system. When users are vicariously immersed in VEs, the perception of texture is scaled in accordance with the body size of the avatar. On the other hand, if users are never immersed, then tactile sensation is identical to the one they are really touching.

The participant's task in this demonstration is to identify the ground texture in VE that matches the tactile sensation with key operation (Fig. 2).

References

1. Kim, J., Gračanin, D., Matkovic, K., Quek, F.: Finger walking in place (FWIP): a traveling technique in virtual environments. In: 2008 International Symposium on Smart Graphics, pp. 58–69. Springer, Heidelberg (2008)
2. Sugiura, Y., Kakehi, G., Withana, A., Fernando, C., Sakamoto, D., Inami, M., and Igarashi, T.: Walky: an operating method for a bipedal walking robot for entertainment. In: ACM SIGGRAPH ASIA 2009 Art Gallery and Emerging Technologies: Adaptation, SIGGRAPH ASIA'09, pp. 79–79. ACM, New York (2009)

Change in the Amount Poured as a Result of Vibration When Pouring a Liquid

Sakiko Ikeno, Ryo Watanabe, Ryuta Okazaki, Taku Hachisu, Michi Sato and Hiroyuki Kajimoto

Abstract Visual and tactile stimulation is known to affect the experience of eating and drinking. In this study, we focused on the vibration of a Japanese sake bottle when used to pour liquid. We manufactured a device that can be attached to the neck of any plastic bottle and investigated how beverage consumption was affected by the vibration. We found that presentation of the vibration affected the amount of poured beverage when visual and sound cues were masked.

Keywords Pouring water · Tactile display · Drink consumption · Vibration · Human food interaction

---

S. Ikeno (✉) · R. Watanabe · R. Okazaki · T. Hachisu · M. Sato · H. Kajimoto
The University of Electro-Communications, Chofu, Japan
e-mail: ikeno@kaji-lab.jp

R. Watanabe
e-mail: r.watanabe@kaji-lab.jp

R. Okazaki
e-mail: okazaki@kaji-lab.jp

T. Hachisu
e-mail: hachisu@kaji-lab.jp

M. Sato
e-mail: michi@kaji-lab.jp

H. Kajimoto
e-mail: kajimoto@kaji-lab.jp

R. Watanabe · R. Okazaki · T. Hachisu
JSPS, Tokyo, Japan

H. Kajimoto
Japan Science and Technology Agency, Tokyo, Japan

H. Kajimoto et al. (eds.), *Haptic Interaction*, Lecture Notes
in Electrical Engineering 277, DOI 10.1007/978-4-431-55690-9_2

1 Introduction

Serving containers are known to affect the experience of eating and drinking, as well as the appearance and flavor of food and beverages. Wanshink and Ittersum [1] showed that food consumption increased when a larger spoon or dish was used to serve food. Furthermore, consumption of beverages is affected by the size and shape of the cup used to serve them. Sakurai et al. [2] and Suzuki et al. [3] changed the appearance of food and beverage containers and regulated the consumption of food and drink using augmented reality technology.

In this study, we focused on the vibration of liquid being poured from a Japanese sake bottle as an audio-haptic rendition of the liquid. Sake bottles are known for their unique "glug" sound and vibration. We believe that these sounds and vibrations affect the subjective impressions of the amount of liquid in the bottle. Thus, we developed a device that can reproduce this vibration [4]. In this study, we manufactured a vibration device that can be attached to the neck of any plastic bottle and investigated how beverage consumption was affected by the subsequent vibrations.

2 An Attachment-Type Device

Figure 1 shows our device, which can be attached to the neck of any plastic bottle. The device has an accelerometer (KXM52-1050, Kionix) and a vibrator (Haptuator Mark II, Tactile Labs Inc.). We used the cap of the plastic bottle as an attachment point for the device. The device does not obstruct the movements required to pour liquid from the bottle.

In our previous report, we reproduced the wave of the vibration using the following model, which comprised two decaying sinusoidal waves with different frequencies [4]:

Fig. 1 An attachment-type device

$$Q(\theta,t) \fallingdotseq \sum_{n=1}^{2} A(\theta)_n \exp(-B(\theta)_n t)\sin(2\pi f(\theta)_n t) \tag{1}$$

where A_n is the wave amplitude, B_n is the attenuation coefficient, f_n is the frequency of the sinusoidal wave, and t is the time period of one wave. Typically, f_1 was around 250 Hz, and f_2 was around 40 Hz.

We used this model to describe the motion of the device. When a user tipped the plastic bottle, the device presented a vibration according to the angle of the bottle.

3 Experiment

We conducted an experiment to see whether the added vibration had an effect on the amount of liquid poured.

3.1 Experimental Environment and Procedure

We recruited four participants (two males and two females, 22–24 years of age). The participants were blindfolded and any sound cues were masked by white noise that was presented through headphones.

Figure 2 shows experimental setting. We prepared 500 ml of water in a plastic bottle with the device attached. There were two conditions: with vibration and without vibration. The participants were asked to pour a designated amount of water, 100, 200, 300 ml, or 400 ml, into a 500 ml cup. For each trial, the participants were shown a marker inside the cup that indicated the level of the designated amount of water. They were then blindfolded and asked to pour the water from the

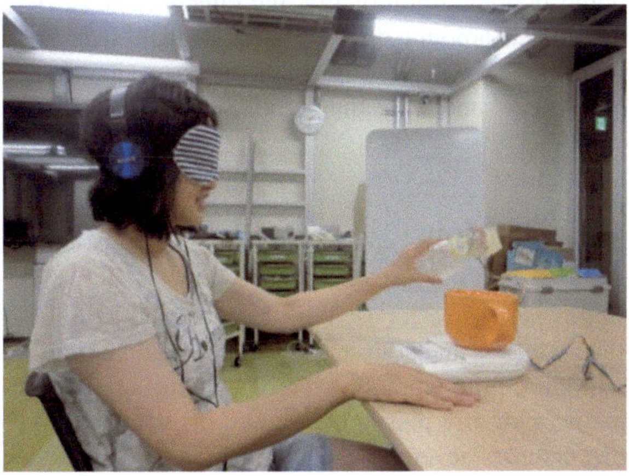

Fig. 2 Experimental setting

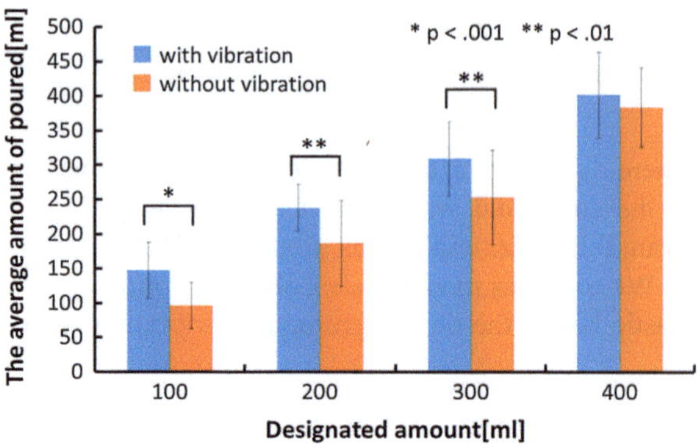

Fig. 3 Average amount of poured

bottle. After they felt that they had poured the designated amount, we measured the amount using an electric scale. There were ten trials for each amount of liquid, 40 trials in total for each participant.

3.2 Results and Discussion

Figure 3 illustrates the average amount of liquid poured and the standard errors in each experimental condition. The horizontal axis represents the designated amount, and the vertical axis represents the average amount of poured water. The participants poured 35 % less in the trials with vibration compared with the trials without vibration at 100 ml, 22 % at 200 ml, 18 % at 300 ml, and 5 % at 400 ml.

We conducted a t-test that revealed significant differences between the pouring conditions at 100 ml ($p = 0.0001$), 200 ml ($p = 0.003$), and 300 ml ($p = 0.006$), although we found no significant differences between the conditions for the 400 ml trials ($p = 0.34$).

These results suggest that the presentation of the vibration affected the amount of liquid poured. Additionally, the effect of the device was more powerful when the designated amount of water was smaller. It is possible that when the amount of liquid to be poured was small, the participants became more conscious of their movements and were thus more affected by the vibration.

4 Conclusions

In this study, we focused on the vibration of liquid being poured from a Japanese sake bottle as an audio-haptic rendition of the liquid. We manufactured a device that reproduced the vibration created by pouring a liquid. This device can be attached to the neck of any plastic bottle. We investigated how pouring of liquids, which is a precursor to beverage consumption, would be affected by the vibration and found that vibration significantly reduced the amount of liquid poured when visual and sound cues were masked.

Future research could test the effect of our device in a real environment with naturalistic visual and sound cues. We also planned to investigate changes in the consumption of beverages associated with the use of our device.

References

1. Wansink, B., Ittersum, K.V.: Bottoms up! the influence of elongation on pouring and consumption volume. J. Consum. Res. **30**, 455–463 (2003)
2. Sakurai, S., Narumi, T., Ban, Y., Kajinami, T., Tanikawa, T., Hirose, M.: Affecting our perception of satiety by changing the size of virtual dishes displayed with a tabletop display. Virtual Augmented Mixed Reality **8022**, 90–99 (2013)
3. Suzuki, E., Narumi, T., Sakurai, S., Tanikawa, T., Hirose, M.: Illusion Cup: Interactive controlling of beverage consumption based on an illusion of volume perception. In: Proceedings of the 5th Augmented Human International Conference (2014)
4. Ikeno, S., Okazaki, R., Hachisu, T., Sato, M., Kajimoto, H.: Audio-haptic rendering of water being poured from sake bottle. Adv. Comput. Entertainment **8253**, 548–551 (2013)

A Study on Upper-Limb Motor Control Using Mirror Illusion in Bimanual Steering

Yoshihiro Tanaka, Taiji Sakajiri and Akihito Sano

Abstract This paper presents the influence of the gravity on the motor control of the upper limb by using the simulated steering system. The generation of the mirror illusion is tested under the condition that one handgrip is fixed and the other handgrip is turned by using the steering system. A preliminary test shows that participants can perceive the steering as being natural (mirror illusion) when the operated arm is turned against gravity, but they hardly perceive the steering as being natural when the operated arm is turned with gravity. This result indicates that the motor control of moving with gravity may be force-oriented and that against gravity may be position-oriented.

Keywords Motor control · Mirror illusion · Upper limb · Gravity · Steering

1 Introduction

Product development increasingly attempts to provide appeal and satisfaction superior to those of competing products. Haptic feel is an important factor. However, haptic feel design has currently a matter of trial and error. Toward the systematic haptic feel design, it is important to reveal the principles of haptic perception. This paper has focused on the steering feel. We developed a simulated steering system that permits independent left- and right-hand steering and torque presentations on a single axis, as shown in Fig. 1. We have investigated the relationship between gravitational and arm movement direction in the perception in bimanual steering [1]. The experimental results demonstrated that the arm moving

Y. Tanaka (✉) · T. Sakajiri · A. Sano
Nagoya Institute of Technology, Gokiso-cho, Showa-ku, Nagoya 466-8555, Japan
e-mail: tanaka.yoshihiro@nitech.ac.jp

T. Sakajiri
e-mail: t.sakajiri.748@nitech.jp

A. Sano
e-mail: sano@nitech.ac.jp

H. Kajimoto et al. (eds.), *Haptic Interaction*, Lecture Notes in Electrical Engineering 277, DOI 10.1007/978-4-431-55690-9_3

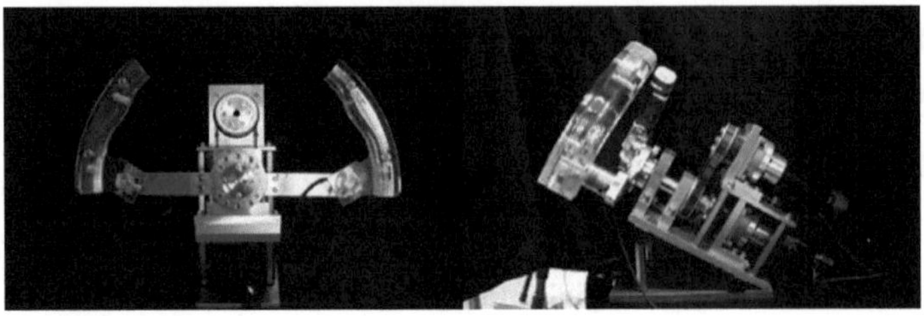

Fig. 1 Simulated steering system

with gravity exerts a larger steering force than the one moving against gravity, and the force exerted in arm movements with gravity is perceived as being smaller than the force exerted against gravity.

In this paper, the relationship between gravitational and arm movement direction in the motor control in bimanual steering is investigated through the mirror illusion using the steering system. The establishment of the mirror illusion is tested under the condition that one hand is fixed in the steering operation. The results show the influence of gravity on the motor control of the upper limb.

2 Experimental Setup and Procedure

The simulated steering system [1] was modified for the experiment, as shown in Fig. 2. For each handgrip, the torque can be presented by two servomotors (maxon motor, RE40) attached at one end of the main driveshaft. The curve of the steering torque presented was based on the prior study of on-center feel [2]. In this experiment, a mirror was set on the center of the steering wheel and one handgrip was fixed. Participants held the fixed handgrip and turned the other handgrip, looking at the mirror where the operated handgrip and arm were mirrored. Here, they could not see the fixed handgrip and their arms. Participants had to answer whether they perceived the steering as being natural on both hands similar to a feel in a normal bimanual steering operation.

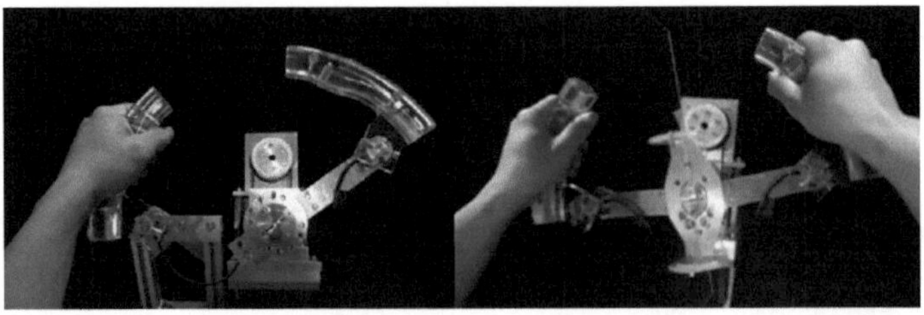

Fig. 2 Experimental setup and operation

3 Results and Discussion

The preliminary experimental result shows that many participants perceived the steering as being natural on the both hands when moving the operated hand against gravity, but did not perceive the steering as being natural when moving it with gravity. In the current experiment, the fixed arm was allowed to exert the steering force, but did not allow to move. The haptic movement perception can be covered by visual information under the mirror illusion. Therefore, the non-generation of the natural steering feel indicates the necessity of the haptic movement perception on the fixed arm. The result indicates that the motor control of moving with gravity may be force-oriented (haptic movement perception is not necessary) and that against gravity may be position-oriented. Furthermore, the additional experiment under the condition that the fixed handgrip was changed to be free, which allowed participants to move the arm but did not allow to exert the steering force, was conducted, and the result showed the opposite tendency in the arm movement direction. This result is consistent with the above-mentioned discussion. Humans may use different motor control depending on gravity. In future, we will investigate the exerted force and movement performance.

References

1. Sakajiri, T., Tanaka, Y., Sano, A.: Relation between gravitational and arm movement direction in the mechanism of perception in bimanual steering. Exp. Brain Res. **231**, 129–138 (2013)
2. Yamada, D., Kuroyanagi, H., Ono, E., Kushiro, I., Konomi, K., Sato, S.: Analysis of on-center steering feel by a steering simulator. Trans. JSAE **40**, 239–244 (2009)

Vibrotactile Cueing for Biasing Perceived Inertia of Gripped Object

Hikaru Nagano, Shogo Okamoto and Yoji Yamada

Abstract Motion-synchronized vibrotactile stimuli on a finger pad influence the perception of the inertia and viscosity of an object being jiggled by the finger (Minamizawa et al. 2007 [1]). We designed a handheld device that imposed vibrotactile stimuli to gripping finger pads. The device allowed us to experience the illusory change in the perceived inertia of the device by vibrotactile stimulation synchronized with the acceleration of the hand movement.

Keywords Vibrotactile stimulation · Mass · Skin stretch

1 Introduction

The perception of the dynamic properties of objects is instrumental for a variety of manipulation tasks. Such percepts are mainly mediated by proprioceptive sensations. In addition, cutaneous sensations influence the perception of objects' properties. For example, the stretching of finger pad skin while handing objects would affect the perception of the mass [2] or frictional properties [3] of objects.

Okamoto et al. [1] demonstrated that the perception of inertia or viscosity of an object was biased by vibrotactile stimuli. In their experiments, the vibrotactile stimuli were provided to a finger pad while jiggling a computer-controlled slider. On the basis of this finding, we designed a handheld device that provided a pseudo-increase in the inertial sensation using vibrotactile stimuli. Such a handheld pseudo-force feedback device is invaluable for mobile interfaces that cannot be grounded to receive the reaction force provided by the device.

H. Nagano (✉) · S. Okamoto · Y. Yamada
Department of Mechanical Science and Engineering, Nagoya University, Nagoya, Japan
e-mail: nagano.hikaru@h.mbox.nagoya-u.ac.jp

H. Kajimoto et al. (eds.), *Haptic Interaction*, Lecture Notes
in Electrical Engineering 277, DOI 10.1007/978-4-431-55690-9_4

2 Illusion of Inertia or Viscosity by Vibrotactile Stimuli in Synchronicity with Hand Motions

The illusory change in the perceived inertia or viscosity of a vibrotactile feedback device has been demonstrated [1]. The phase of the change in stretched finger skin caused by an increase in the mass of an object is close to the phase of the acceleration of the finger that jiggles the object. On the basis of this relationship, the perception of the inertia of the object was influenced by the vibrotactile stimuli on the finger pad that is synchronized with the finger acceleration. As shown in Fig. 1, the illusory change in the perception of the inertia of an object was presumed to be caused by the superactivation of skin mechanoreceptors by vibrations. The vibrotactile stimuli would add to the receptors' activation in synchronizing the stretching of finger skin. In addition, the vibrotactile stimuli that are synchronized with the finger velocity biased the perceived viscosity of an object on the basis of the relationship that the phase of the stretched skin change due to an increase in the viscosity of an object was close to that of the finger velocity.

3 Apparatus and Stimuli

The device (Fig. 2) we designed was composed of a box-shaped acrylic enclosure, two actuators (Force Reactor, Alps Electric Co., Tokyo, Japan), a microcontroller (mbed NXP LPC1768, NXP semiconductors, the Netherlands), an accelerometer (ADXL335, Analog Devices, Norwood, USA), an electronic circuit, and a power supply. All components were internally located in the enclosure. Each of the two actuators was attached to the ceiling and bottom of the enclosure, which generated vibrotactile stimulation as commanded by the microcontroller. The rates of control and measurement were set to 1 kHz.

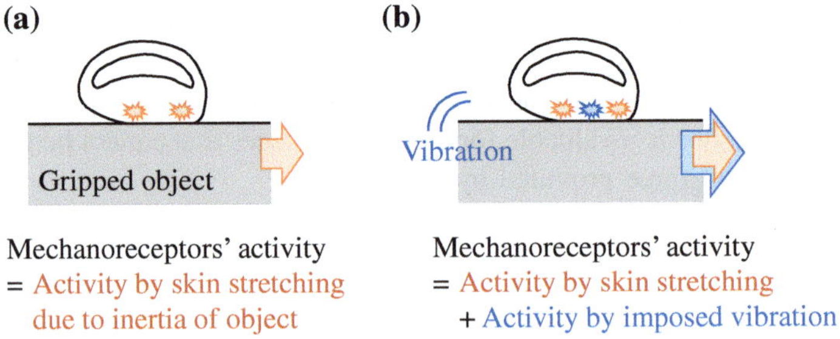

Fig. 1 Imposed vibration biases the perception of inertia of an object. **a** Without vibration. **b** With imposed vibration

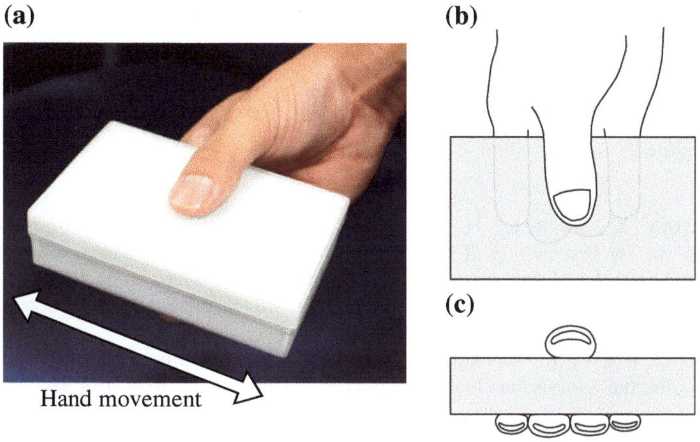

Fig. 2 Apparatus used to apply vibrotactile stimulation to multiple fingers. **a** An apparatus grasped by five digits. **b** Top view. **c** Side view

The actuators were driven by PWM at a frequency of 280 Hz through a current amplifier. The fast adaptive units could be mainly activated by the vibrations. The duty ratio was determined by

$$D(t) = \begin{cases} a_i & (|\ddot{x}(t)| < b_i) \\ a_i\left(1 + \frac{(|\ddot{x}(t)| - b_i)}{c_i}\right) & (|\ddot{x}(t)| \geq b_i), \end{cases} \tag{1}$$

where $\ddot{x}(t)$, a_i, b_i, and c_i are the acceleration of the device, the duty ratio when the finger movement was at rest, a constant value representing the insensitive range of finger acceleration, and the gain of the acceleration of the box, respectively; $a_i = 0.02$, $b_i = 15.0\,\text{m/s}^2$, and $c_i = 2.0\,\text{m/s}^2$. In our experiment, this combination of values was likely to bias the perceived inertia of the apparatus.

4 Demonstration

In order to authenticate the illusion, voluntary participants swung the device horizontally. To control the conditions of the demonstration, in which the frequency and width of the horizontal movements were 1.75 Hz and 0.5 m, respectively, the participants swung the device in rhythm with a metronome with the instructed width as accurately as possible. The frequency and width of the movements were determined by the researcher. The participants experienced illusory and non-illusory conditions and reported the increases in inertia under the illusionary conditions.

Ackowledgement This work was in part supported by MEXT KAKENHI Shitsukan (25135717, 23135514) and 13J02247.

References

1. Minamizawa, K., Kajimoto, H., Kawakami, N., Tachi, S.: Wearable to present gravity sensation. In: The Proceedings of the Second Joint EuroHaptics Conference and Symposium on Haptic Interfaces for Virtual Environment and Teleoperator Systems, pp. 133–138 (2007)
2. Provancher, W.R., Sylvester, N.D.: Fingerpad skin stretch increases the perception of virtual friction. IEEE Trans. Haptics **2**(4), 212–223 (2009)
3. Okamoto, S., Konyo, M., Tadokoro, S.: Vibrotactile stimuli applied to finger pads as biases for perceived inertial and viscous loads. IEEE Trans. Haptics **4**(4), 307–315 (2011)

Auditory Feedback for Ear Picks

Masahiro Koge, Yosuke Kurihara, Ryuta Okazaki, Taku Hachisu and Hiroyuki Kajimoto

Abstract We often clean others' ears for the purpose of hygiene and communication. However, this activity has a risk of injuring the ears from applying too much force because it is difficult to grasp the movement and position of an ear pick. To solve this problem, we present novel techniques to provide cues for grasping behavior of the ear pick using auditory feedback. We implemented two techniques: (1) direct feedback of scratch sound and (2) conversion of force applied to the ear canal to audible signal. We conducted two experiments to study whether these techniques can help users control the exerted force. Contrary to our expectation, the results of the first experiment showed that the direct feedback of scratch sound had no helpful effect on force control. However, the results of the second experiment showed the marginally significant effect that the conversion of force applied to the ear canal to audible signal reduced force. This result indicates that the audification of the force helps users to control the force.

Keywords Auditory feedback · Human interface · Ear cleaning · Ear pick

M. Koge (✉) · Y. Kurihara · R. Okazaki · T. Hachisu · H. Kajimoto
The University of Electro-Communications, Chōfu, Japan
e-mail: koge@kaji-lab.jp

Y. Kurihara
e-mail: kurihara@kaji-lab.jp

R. Okazaki
e-mail: okazaki@kaji-lab.jp

T. Hachisu
e-mail: hachisu@kaji-lab.jp

H. Kajimoto
e-mail: kajimoto@kaji-lab.jp

R. Okazaki · T. Hachisu
JSPS, Tokyo, Japan

H. Kajimoto
Japan Science and Technology Agency, Tokyo, Japan

H. Kajimoto et al. (eds.), *Haptic Interaction*, Lecture Notes
in Electrical Engineering 277, DOI 10.1007/978-4-431-55690-9_5

1 Introduction

Ear cleaning is a familiar behavior for many people and it is performed often. The main purpose of ear cleaning is to remove the earwax in the ear canal. However, there are many reasons why we may clean others' ears, such as cleaning one's partner's ears or parents cleaning their infants' ears; thus, ear cleaning is also a type of communication, if not an entertainment.

However, cleaning the ears of others is generally difficult because the width of the ear canal is very small—6 mm on average in an adult—and its depth is 30 mm on average [6]. Furthermore, while tactile cues from the ear canal play an important role when cleaning one's own ears, this cue is removed when cleaning other people's ears, and the small tactile signal transmitted from the ear pick becomes the only cue. These difficulties can result in excessive force and sometimes injury.

As a means for solving this problem, ear picks with visual assistance functionality are commercially available. Some illuminate the internal ear via LEDs, while others use an endoscope. However, two issues remain to be solved: occlusion by the ear pick itself and cost.

In this paper, we propose a method to improve the safety of ear cleaning by supplementing auditory cues. We evaluated two methods: the direct feedback of scratch sound and the conversion of force applied to the ear canal into an audible signal.

2 Related Work

Attempts to improve the operability of the tool by applying tactile or aural cues have principally been made in the medical field. Yao et al. [7] developed an enhanced probe that can detect small wounds used in arthroscopy. The probe detects a slight acceleration signal that occurs when it traces damage, and magnifies it for tactile and auditory sensation. Also, in the field of telesurgery, detection of the acceleration signal at the slave side is magnified and transmitted to the master side as tactile and auditory feedback [4, 5]. These studies reported that the operability of the tool was improved by the auditory cues.

3 Method 1: Direct Feedback of Scratch Sound

System: We developed an ear pick device that detects and amplifies scratch sound [3]. Figure 1 shows the system overview. The device is composed of a silicon microphone (Knowles Electronics Inc., SPU0409HD5H), an ear pick made of bamboo, an amplifier, and an earphone. The scratch sound from the ear pick is

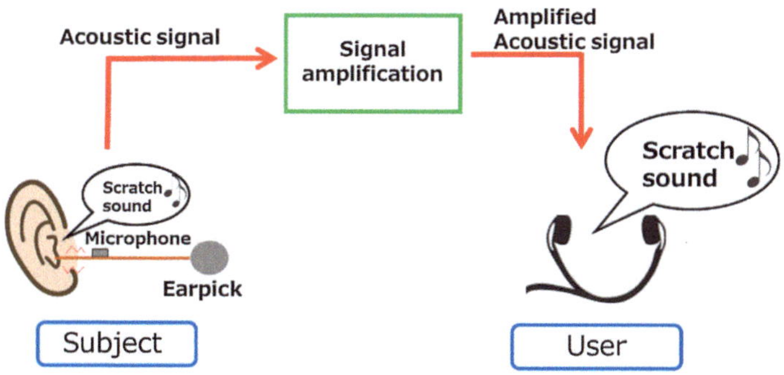

Fig. 1 Direct feedback of scratch sound

detected by the microphone, amplified by the amplifier, and presented to the user via the earphone. Users can adjust the amplitude by changing the volume of the amplifier.

3.1 Experiment

The purpose of the experiment was to determine whether users can more clearly understand the contact state of the ear pick using the scratch sound. We expected that this feedback would facilitate more precise force control, so that the force exerted while cleaning would be changed.

Experiment environment: Fig. 2 shows the experimental system overview. The system comprises the ear model for medical procedure practice (Kyoto Kagaku Co., Ltd., foreign body removal practice unit, 11222-000[M88]) and a six-axis force sensor (NITTA Corp., TFS12-10); both were fixed to a desk with an acrylic plate. We also used pseudo earwax (Kyoto Kagaku Co., Ltd., foreign materials for

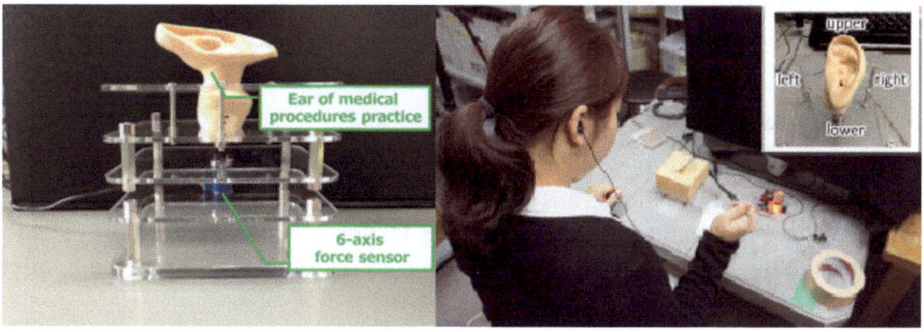

Fig. 2 Artificial ear and the 6-axis force sensor for the experiment (*left*) and overview of the experiment (*right*)

removal practice) for the experiment. The force sensor was situated 50 mm under the ear model and detected the torque applied to the ear canal. The signal of the force sensor was recorded by the PC via an AD board (interface Inc., PCI-3523A, voltage range ±10 V, 12-bit resolution).

Participants: We recruited ten participants (seven males, three females, 22–29 years old, all right-handed). None reported any auditory impairments.

Experimental procedure: Participants sat in front of the desk and wore a single earphone in their right ear (Fig. 2). They controlled the volume to the extent that they could hear sound clearly, but it was not unpleasant. They were asked to treat the pseudo ear like the actual ear and not to touch the acrylic plate. They were informed that the pseudo earwax was adhered evenly inside the pseudo ear, and they were instructed to clean separate regions one by one, as shown in Fig. 2. Ear cleaning was carried out until participants felt that it was cleaned. Participants were asked to take 20-s rest after one region was cleaned. The experimenter checked that the pseudo earwax had been removed completely after ear cleaning. Participants carried out the task under two conditions: with and without scratch sound feedback. To avoid order effects, we separated the participants into two groups. One group started in the with-sound-feedback condition (participants A, B, C, D, and E) and the other group started in the without-sound-feedback condition (participants F, G, H, I, and J). We recorded the maximum norm of the torque applied to each area as the evaluation value for each condition. After the experiment, we conducted a questionnaire on the usability of the system.

3.2 Results

Figure 4 (left) shows the results of this experiment. The horizontal axis shows participants, and the vertical axis shows the average of the maximum norm of the torque applied to each area of the ear canal. Error bars show standard deviation. A one-way analysis of variance (ANOVA) of the three conditions, feedback order, feedback conditions, and areas, did not show any significant difference in the mutual effect and main effect.

3.3 Discussion

Contrary to our expectations, this result showed that scratch sound feedback does not facilitate more precise force control. This feedback is considered to be effective in understanding the texture and state of touch [6, 7], but it was not found to affect force control.

4 Method 2: Conversion of Force Applied to the Ear Canal to Audible Signal

Based on the results of the previous section that scratch sound feedback does not affect force control, we propose a new method that converts force applied to the ear canal into an audible signal. There are several studies that have used audio signals for biofeedback, such as to improve body balance and motor skills [1, 2]. We hypothesized that by converting force applied to the ear canal to an audible signal, users would clearly be able to grasp the amount of exerted force and thus would better be able to control it.

System: The system setup was similar to the previous experiment, but a 6-axis force sensor was situated 65 mm under the ear model. Figure 3 shows the system structure. The force applied to the pseudo ear was obtained from the 6-axis force sensor and sent to the PC via the AD board (Interface Inc., PCI-3523A, voltage range ±10 V, 12-bit resolution). The norm of the torque was calculated by the PC, multiplied by a 1.5 kHz sine wave, and presented to the user via earphone. In this way, users heard a stronger tone when they exerted a stronger force. In future systems, the force will be measured by a force sensor in the ear pick, e.g., by piezoelectric film, but here, we used an external sensor as an initial trial. In our system, the maximum volume was set to 97 dB (A) when the norm of the torque was 3.0 N cm.

4.1 Experiment

Participants: We recruited six participants (three males, three females, 22–29 years old, all right-handed). None reported any auditory impairments.

Experimental procedure: Participants sat in front of the desk and wore headphones in both ears (Fig. 3). They were asked to treat the pseudo ear like the actual ear

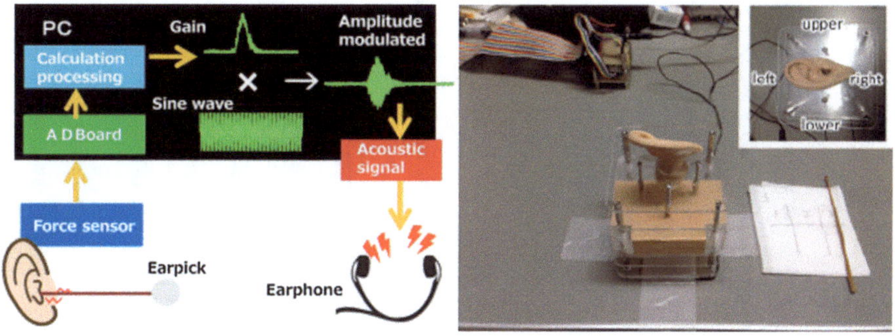

Fig. 3 System whereby force applied to the ear canal was converted to an audible signal (*left*) and overview of the experiment (*right*)

and not to touch the acrylic plate. They were informed that the pseudo earwax was adhered to the upper area of the pseudo ear, and they performed ear cleaning. In this experiment, we fixed the number of scratches as three times in one trial and asked the participants to clean as best as possible within this number. When one set was finished, the experimenter checked that the pseudo earwax had been removed completely, and set new pseudo earwax at the same area. Participants carried out the trials under two conditions: with and without audible feedback of the force. We divided the participants into two groups. One group started in the with-audible-feedback condition (participants A, B, and C) and the other group started in the without-audible-feedback condition (participants D, E, and F). Each condition had five trials, and the participants performed 10 trials in total. The first trial in each condition was regarded as a practice and thus not included in the evaluation. We recorded the maximum norm of the torque applied to each area. After the experiment, we conducted a questionnaire survey on the usability of the system.

4.2 Results

In this experiment, we fixed the number of scratches as three times in one trial, but we used the results from the second scratch because it was more suitable than the other two. Figure 4 (right) shows the result of the experiment. The horizontal axis shows participants, and the vertical axis shows the average of the maximum norm of the torque applied to the ear canal. Error bars show standard deviation. A one-way analysis of variance (ANOVA) of the two conditions, feedback order and feedback condition, showed a marginally significant difference in feedback condition ($F(1, 4) = 4.91$, $p < 0.1$).

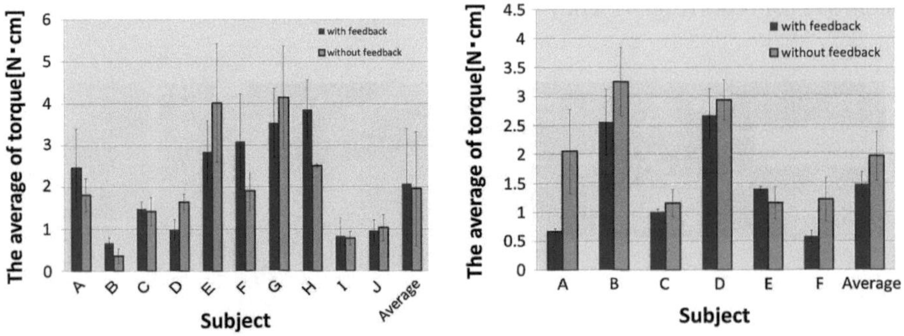

Fig. 4 Results of the experiment. Method 1 (*left*) and method 2 (*right*)

4.3 Discussion

As we expected, this result showed that an aural signal converted from force reduces exerted force, even though we did not tell participants what they were expected to do. Figure 4 shows that the standard deviation of the maximum norm of the torque became small, which suggests that the exerted force was stabilized by this feedback.

5 Conclusion

In this paper, we proposed and evaluated two methods to improve the safety of ear cleaning by supplementing auditory cues. The first method was direct feedback of scratch sound from the ear pick when cleaning another's ears. The results indicated that this sound cue did not effectively facilitate the user's force control. The second method was to convert the force applied to the ear canal into an audible signal. The results showed that the force applied to the ear canal was decreased and stabilized.

Although the system was implemented in the experimental environment, we think that it would be quite easy to make the system compact, such as by using a small strain gauge.

References

1. Chiari, L., Dozza, M., Cappello, A., Horak, F.B., Macellari, V., Giansanti, D.: Audio-biofeedback for balance improvement: an accelerometry-based system. IEEE Trans. Biomed. Eng. **52**(12), 2108 (2005)
2. Hasegawa, S., Ishijima, S., Kato, F., Mitake, H., Sato, M.: Realtime sonification of the center of gravity for skiing. In: AH'12 Proceedings of the 3rd Augmented Human International Conference (2012)
3. Koge, M., Kurihara, Y., Okazaki, R., Hachisu, T., Kajimoto, H.: Safer ear cleaning by adding auditory feedback. Entertain. Comput. (2013) (In Japanese)
4. McMahan, W., Gewirtz, J., Standish, D., Martin, P., Kunkel, J.A., Lilavois, M., Wedmid, A., Lee, A.I., Kuchenbecker, K.J.: Tool contact acceleration feedback for telerobotic surgery. IEEE Trans. Haptics **4**(3), 210–220 (2011)
5. Okamura, A.M.: Methods for haptic feedback in teleoperated robot-assisted surgery. Ind. Rob. Int. J. **31**(6), 499–508 (2004)
6. Schuenke, M., Schulte, E., Schumacher, U., Ross, L.M., Lamperti, E.D., Voll, M.: THIEME Atlas of Anatomy Image Collection—Head and Neuroanatomy, pp. 140–143. Thieme Medical Publishers, Stuttgart (2007)
7. Yao, H.-Y., Hayward, V., Ellis, R.E.: A tactile enhancement instrument for minimally invasive surgery. Comput. Aided Surg. **10**(4), 233–239 (2005)

Visual Stimulation Influences on the Position of Vibrotactile Perception

Arinobu Niijima and Takefumi Ogawa

Abstract Tactile feedback is important for interaction in virtual reality. Some previous works employed vibration motors for tactile display. However, it is difficult to recognize the position of vibrotactile perception because the resolution of tactile perception is not so high. Our previous works and other previous works describe that visual stimulation seemed to influence tactile perception. In this paper, we describe some experiments to verify the spatial resolution of vibrotactile perception on a forearm and a palm and to investigate the influence of visual stimulation on vibrotactile perception. The results show that visual stimulation influences the position of vibrotactile perception. From this perspective, we propose a vibrotactile display with LEDs for changing the position of vibrotactile perception.

Keywords Cross-modal · Vibration motor · Visual stimulation · Vibrotactile perception · Phantom sensation

1 Introduction

Tactile feedback is important for interaction in virtual reality [1, 2]. However, human tactile sensation is not so clear as visual sensation [3]. Therefore, it is difficult to clearly recognize where it is stimulated. Some tactile displays use vibration motors [4–6]. Most of them actuate less than three motors simultaneously. If more than four separate motors are vibrated simultaneously, users will be confused and not recognize where it is stimulated.

A. Niijima (✉) · T. Ogawa
The University of Tokyo, Bunkyō, Japan
e-mail: a.niijima@ogawa-lab.org

T. Ogawa
e-mail: ogawa@nc.u-tokyo.ac.jp
URL: http://www.ogawa-lab.org

H. Kajimoto et al. (eds.), *Haptic Interaction*, Lecture Notes
in Electrical Engineering 277, DOI 10.1007/978-4-431-55690-9_6

In our study, we have previously observed that the positions of vibrotactile perception shift to those of visual stimulation by presenting visual and tactile stimulation in different places on a forearm [7]. Some studies show the cross-modal between visual and tactile sensations [8–10]. From these results, we consider that visual stimulation makes vibrotactile perception clear. In this paper, we conducted some experiments to verify the spatial resolution of vibrotactile perception on a forearm and a palm and to investigate the influence of visual stimulation on vibrotactile perception.

2 Experiment

We conducted two experiments. Experiment 1 measured spatial resolution of vibrotactile perception on a forearm and a palm. Experiment 2 investigated influence of visual stimulation on vibrotactile perception.

The experimental environment is shown in Fig. 1. We used ARToolkit for presenting a virtual ball as visual stimulation [11] and coin-type vibration motors (VMT-003, Yatsugatake Club) whose radius is 5 mm for tactile stimulation. Arduino was used to control the power of the motors [12]. Subjects wore HMD (Vuzix Wrap 920) to look at their own left forearm and palm.

2.1 Experiment 1: Spatial Resolution of Vibrotactile Perception

2.1.1 Method

Seven subjects (24–37 years old; six males, one female) participated in this experiment. They were equipped with nine vibration motors on the left forearm and palm as shown in Fig. 2. The distance of each motor was 20 mm. The motors were

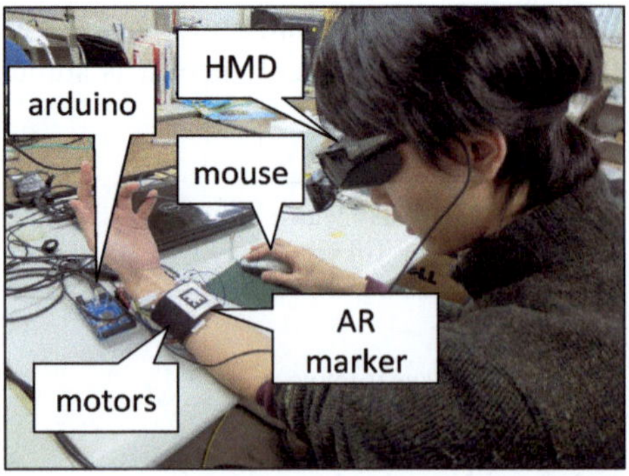

Fig. 1 Experimental environment

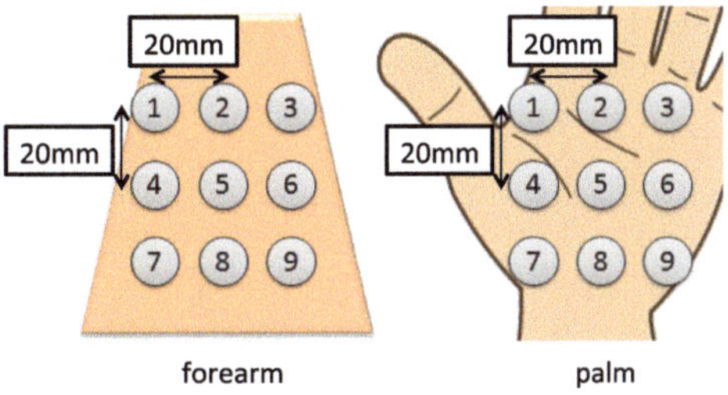

Fig. 2 Layout of motors

supplied with 3.0 V, and the frequency of them was about 200 Hz. In the task, a single motor, which was selected randomly, vibrates, and then, subjects answered which motor vibrates. The number of trials was 100. We calculated the percentage of correct answers.

2.1.2 Results and Discussion

Figures 3 and 4 show the percentage of correct answer in case of a forearm and a palm, respectively. It was different between each vibrating position, and the average of the correct answer rate of palm was higher than the one of forearm. The reason of difference seemed to be the spatial resolution of tactile perception on each body part [3]. Figures 5 and 6 show the frequency distribution of answer in case of a forearm and a palm, respectively. Subjects answered wrong position which was next to the correct position. It seemed to be difficult to recognize the vibrating position clearly.

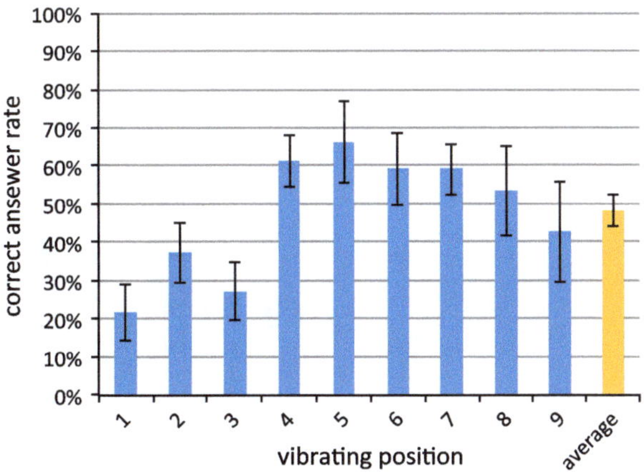

Fig. 3 Correct answer rate on a forearm

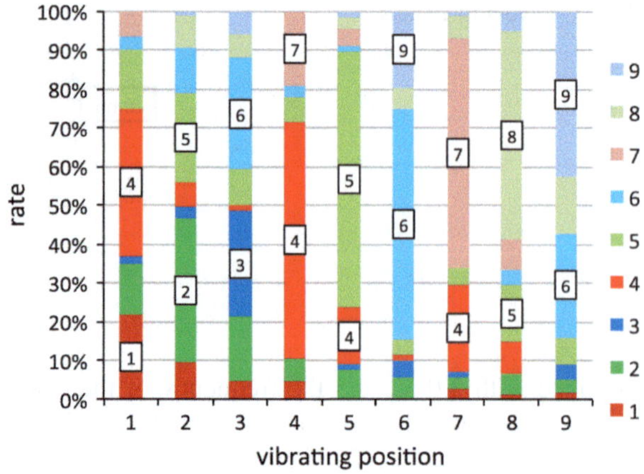

Fig. 4 Frequency distribution of answer on a forearm

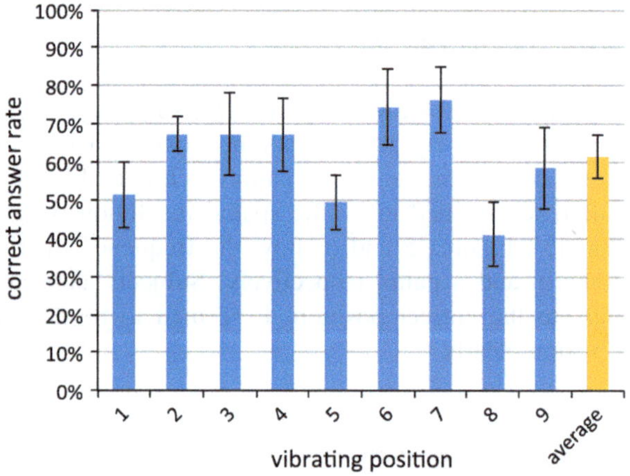

Fig. 5 Correct answer rate on a palm

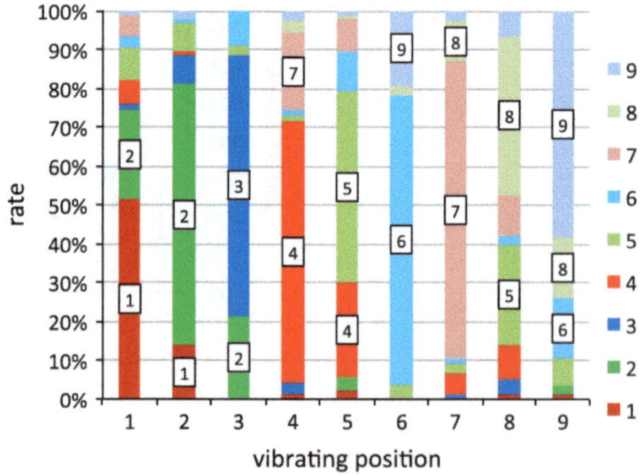

Fig. 6 Frequency distribution of answer on a palm

2.2 *Experiment 2: Influence of Visual Stimulation on Vibrotactile Perception*

2.2.1 Method

Six subjects (24–37 years old; five males, one female) participated in this experiment. They were equipped with nine vibration motors on the left palm as Experiment 1. The vibration pattern was the same as Experiment 1. A virtual ball was presented simultaneously as visual stimulation. The radius of the ball was 5 mm. The position of the ball was selected randomly from 1 to 9 in Fig. 2. The position of the ball was not always the same as the vibrating position. We did not inform subjects of the relationship between the vibrating position and the ball position. Subjects answered which motor was vibrating. The number of trials was 200.

2.2.2 Results and Discussion

Figure 7 shows the correct answer rate when the vibrating position and the ball position were the same. When the virtual ball was presented on the same position as the vibrated motor, subjects seemed to recognize clearly where was the vibrating position, so the correct answer rate was higher than the results of Experiment 1 as shown in Fig. 5. It means that the visual stimulation influences vibrotactile perception and it makes vibrotactile perception clear.

Figures 8 and 9 show the rate at which subjects answered that the motor under the virtual ball was vibrating when the virtual ball was presented next to vibration position in a horizontal and vertical direction, respectively. Subjects sometimes perceived vibrotactile stimulation under the virtual ball. It means that the visual stimulation influences on vibrotactile perception. In other words, the visual stimulation has the potential for causing tactile illusions.

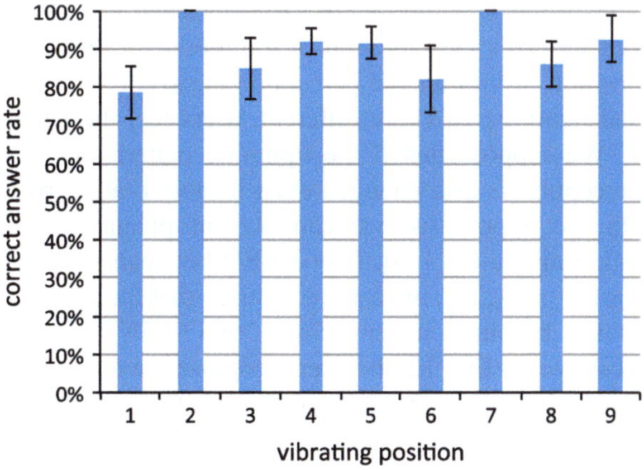

Fig. 7 Correct answer rate on a palm with visual stimulation

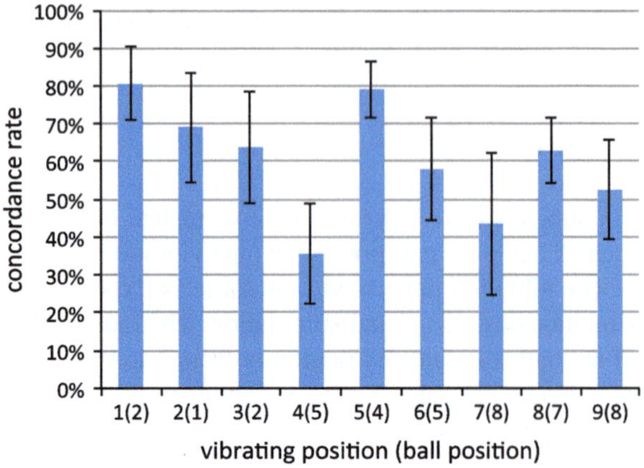

Fig. 8 Concordance rate of vibrotactile perception and visual stimulation when the ball and the motor were distant horizontally

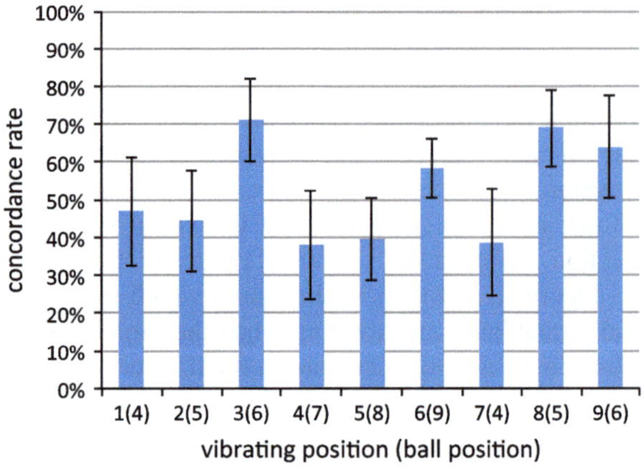

Fig. 9 Concordance rate of vibrotactile perception and visual stimulation when the ball and the motor were distant vertically

From the results, we propose vibrotactile display with LEDs as shown in Fig. 10. The display is composed of LED tapes and vibration motors. If the flash of the LED synchronizes the vibration of the motor on the same position, users recognize the position of vibrotactile perception clearly. Even if the positions of vibrating motor and lightning LED are different, users feel the LED is vibrating.

We suppose that if the lightning pattern of the LEDs is changed, vibrotactile perception is also changed even if the vibrating pattern of the motors is the same. On the hypothesis, we can present phantom sensation ON/OFF by changing lightning pattern [13]. If one LED between separated two motors is turned on as shown in Fig. 11, the user recognizes that the vibrating position is the middle of two motors by phantom sensation. If two LEDs on separated two motors are turned on as shown in Fig. 12, the user recognizes two vibrating positions on the two motors. Furthermore, we can control vibrotactile positions by changing the visual stimulation position.

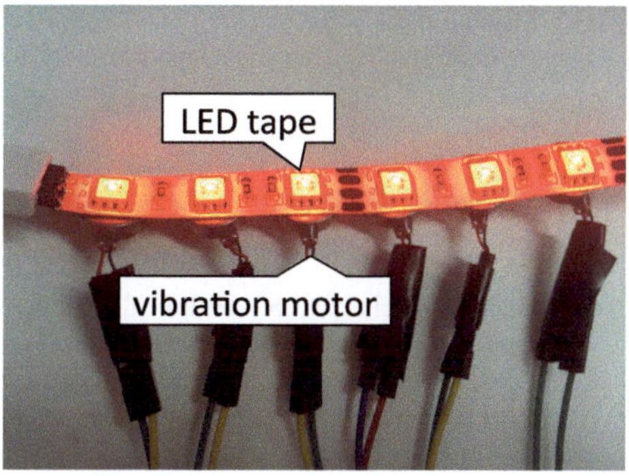

Fig. 10 Vibrotactile display with LED

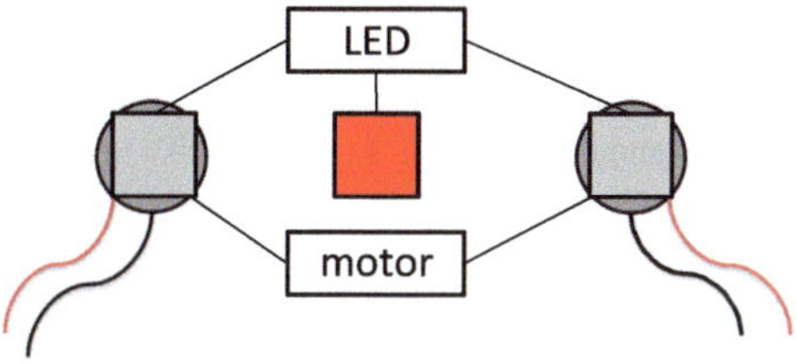

Fig. 11 Phantom sensation ON

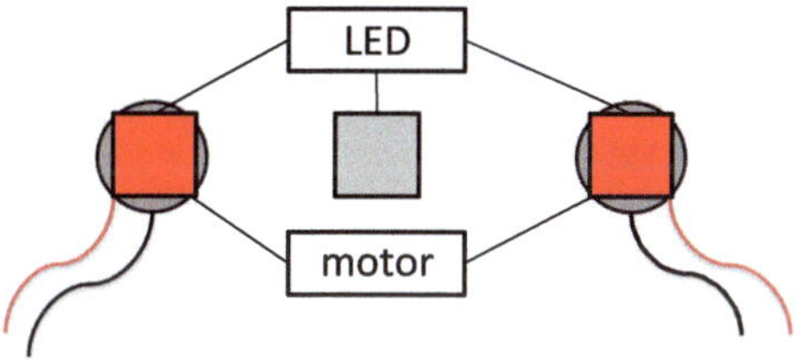

Fig. 12 Phantom sensation OFF

3 Conclusions and Future Works

Tactile feedback is important for virtual reality, but vibrotactile perception is not so clear. We conducted an experiment to verify the resolution of vibrotactile perception on a forearm and a palm. The result shows that it is difficult to recognize the position of vibrotactile perception. We also conducted another experiment to investigate the influence of visual stimulation on vibrotactile perception. The result shows that the visual stimulation influences vibrotactile perception and it makes vibrotactile perception clearly. From these results, we propose vibrotactile display with LEDs as shown in Fig. 10. In the future work, we investigate the resolution of vibrotactile perception with our proposal display.

Acknowledgement This research was supported in part by a Grant-in-Aid for Scientific Research (C) numbered 25330227 by the Japan Society for the Promotion of Science (JSPS).

References

1. Weber, B., Schatzle, S., Hulin, T., Preusche, C., Demi, B.: Evaluation of a vibrotactile feedback device for spatial guidance. In: World Haptics Conference, pp. 349–354 (2011)
2. Sadihov, D., Migge, B., Gassert, R., Kim, Y.: Prototype of a VR upper-limb rehabilitation system enhanced with motion-based tactile feedback. In: World Haptics Conference, pp. 449–454 (2013)
3. Lederman, S.J., Klatzky, R.L.: Haptic perception: a tutorial. Atten. Percept. Psychophys. **71** (7), 1439–1459 (2009)
4. Israr, A., Poupyrev, I.: Tactile brush: drawing on skin with a tactile grid display. In: CHI2011, pp. 2019–2028 (2011)
5. Ueda, S., Uchida, M., Nozawa, A., Ide, H.: A tactile display using phantom sensation with apparent movement together. Electron. Commun. Jpn. **91**(12), 29–38 (2008)
6. Barghout, A., Cha, J., El Saddik, A., Kammerl, J., Steinbach, E.: Spatial resolution of vibrotactile perception on the human forearm when exploiting funneling illusion. In: Haptic Audio Visual Environments and Games, pp. 19–23 (2009)
7. Niijima, A., Ogawa, T.: Influence analysis of visual stimuli on localization of tactile stimuli in augmented reality. In: Proceedings of Virtual Reality, pp. 105–106 (2012)
8. Spence, C., Pavani, F., Driver, J.: Spatial constraints on visual-tactile cross-modal distractor congruency effects. Cognit. Affect. Behav. Neurosci. **4**(2), 148–169 (2004)
9. Craig, J.C.: Visual motion interferes with tactile motion perception. Perception **35**, 351–367 (2006)
10. Sano, Y., Hirano, Y., Kimura, A., Shibata, F., Tamura, H.: Dent-softness illusion in mixed reality space: further experiments and considerations. In: Proceedings of Virtual Reality, pp. 153–154 (2013)
11. NyARToolKit project. http://nyatla.jp/nyartoolkit/wp/. Retrieved 24 July 2014
12. Arduino. http://www.arduino.cc/. Retrieved 24 July 2014
13. Alles, D.S.: Information transmission by phantom sensations. IEEE Trans. Man Mach. Syst. **11**, 85–91 (1970)

Haptic Assistance of Spatial Pointing with Simple Vibrotactile Feedback for Gesture Interfaces

Seonghwan Kim, Masashi Konyo and Satoshi Tadokoro

Abstract Gesture input with body motion has an operating limitation due to a lack of human precise positioning capability of the hand in the air. This study proposes simplified vibrotactile feedback methods to assist spatial pointing performances. The proposed methods demonstrated simplified tactile feedbacks, which were just non-directional vibratory intensity or a few vibration patterns.

Keywords Gesture interface · Tactile feedback · Pointing task

1 Introduction

Gesture interface has begun to be used widely recently. Devices such as Kinect and Leap Motion are being increasingly used in non-contact human-computer interfaces and body position based games. There are many advantages to use gesture interface such as easy operation in 3D space, wide operation range. But, there are many problems such as poor operation feeling for operating erroneous operation.

In this study, we propose 2 methods to reduce these problems with vibrotactile feedback. The first is applying non-directional vibration which is proportional to the distance to the center to guess the position. The second is recognizing buttons by a few vibration patterns less than the number of buttons. It is normally considered the impossible, but it becomes possible by combining the advantage of gesture interfaces.

S. Kim (✉) · M. Konyo · S. Tadokoro
Graduate School of Information Sciences (GSIS), Tohoku University,
6-6-01 Aramaki Aza Aoba, Aoba-ku, Sendai-shi, Miyagi, Japan
e-mail: kimsh@rm.is.tohoku.ac.jp

H. Kajimoto et al. (eds.), *Haptic Interaction*, Lecture Notes
in Electrical Engineering 277, DOI 10.1007/978-4-431-55690-9_7

2 Methods

2.1 Non-directional Vibration Which Is Proportional to the Distance to the Center

We can guess the position of our hand roughly without vision. But we cannot guess the position of our hand in detail without vision. We propose a method to apply non-directional vibration which is proportional to the distance to the center as shown in Fig. 1. It is normally hard to identify the position with non-directional vibration, but in this case, we can guess the position more exactly by the combination of non-directional vibration and the movement of hand.

2.2 Recognizing Buttons by a Few Vibration Pattern Less than the Number of Buttons

We apply a vibrotactile feedback like collision when we touch a button [1]. And we make 3 distinguishable patterns and set in the button as shown in Fig. 2. When you want to push a button, you will push buttons near that button since we can guess the position of our hand roughly without vision. Because of that, we can distinguish contact condition and recognize what button is touched by this method.

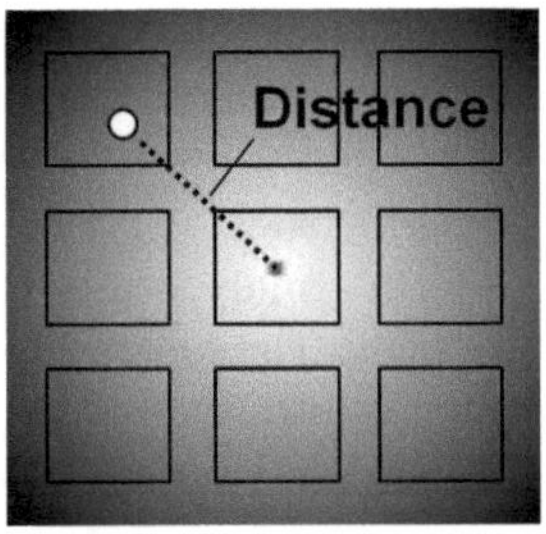

Fig. 1 Vibration strength

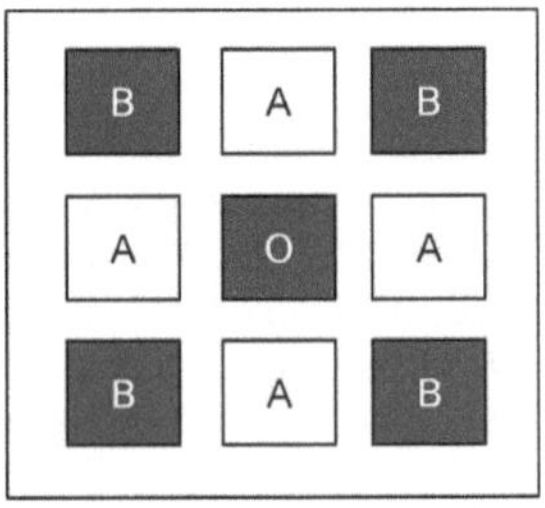

Fig. 2 Vibration pattern

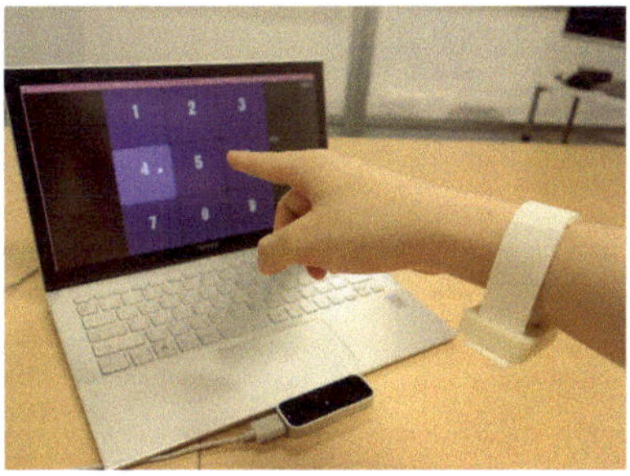

Fig. 3 Overview

3 Demonstration

Demonstrations are performed with vibrator bracelet as shown in Fig. 3.

The following is the flow of button demonstration. First, while moving a hand, you can guess the position of the hand more exactly by comparing the movement of your hand and non-directional vibration which is proportional to the distance to the center. Then, you move your hand and select the button that you want vertically. When you touch a button, you can recognize what you touched by vibration pattern.

We prepare map demonstration as the application of those methods. There are 2 modes. One-finger is moving mode and two-finger is zoom mode. We apply the vibration which is proportional to the amount of change. It is possible to intuitively know the amount of change.

Reference

1. Okamura, A.M., Cutkosky, M.R., Dennerlein, J.T.: Reality-based models for vibration feedback in virtual environments. IEEE/ASME Trans. Mechatron. **6**(3), 245–252 (2001)

Pressure Sensation Elicited by Rapid Temperature Changes

Ryo Watanabe and Hiroyuki Kajimoto

Abstract We found that force or tactile sensation occurred when temperature of thermal elements which were statically touched by subject's skin changes rapidly. This study aims to clarify its mechanism and to investigate its nature. In this paper, we conducted an experiment to verify incidence rate and quality of this thermal-tactile illusion. As result of the experiment, some kind of vertical sensation occurred and interpretations of occurred sensation were various.

Keywords Thermal-tactile illusion · Haptic illusion

1 Introduction

We found that illusional force or tactile sensation occurred when temperature of thermal elements which were statically touched by subject's skin changes rapidly. Illusional upward moving and pressure or force sensation elicited when stimulus temperature rises rapidly. On the other hand, illusional downward moving or pulling sensation elicited when stimulus temperature declines rapidly. This study aims to clarify its mechanism and to investigate its nature. In this paper, we conducted an experiment to verify incidence rate and quality of this thermal-tactile illusion.

R. Watanabe (✉) · H. Kajimoto
The University of Electro-Communications, Chōfu, Japan
e-mail: r.watanabe@kaji-lab.jp

H. Kajimoto
e-mail: kajimoto@kaji-lab.jp

R. Watanabe
JSPS, Tokyo, Japan

H. Kajimoto
Japan Science and Technology Agency, 1-5-1 Chofugaoka, Chofu, Tokyo 182-8585, Japan

H. Kajimoto et al. (eds.), *Haptic Interaction*, Lecture Notes
in Electrical Engineering 277, DOI 10.1007/978-4-431-55690-9_8

2 Experiment

2.1 Experiment System

Thermal stimulator consists of heat sink, fan, thermistors, and Peltier elements (Fig. 1). Applied voltage to Peltier elements was PID controlled by motordriver. Area of presenting thermal stimulus was 40 mm × 80 mm.

2.2 Condition

Six male participants aged between 21 and 25 years participated in the study. The experiment was conducted in a room maintained at 28 °C. Thermal stimulator presented following 3 conditions of thermal stimuli.

(a) Constant 37 °C
(b) Constant 27 °C
(c) 0.25 Hz sine wave and temperature 27–32 °C

Both the temperatures 37 and 27 °C which do not damage the skin have 5 °C difference from skin temperature 32 °C [1].

2.3 Procedure

Participants touched thermal stimulator which presenting one of stimuli condition with right hand and kept their hand statically during 30 s. After 30 s, participants release their hand from the stimulator and answered whether some kind of vertical sensation (moving, tactile, or force) occurred, and what specifically kind of sensation was perceived. All participants participated in 30 trials (10 trials per a stimulus condition).

Fig. 1 Thermal stimulator

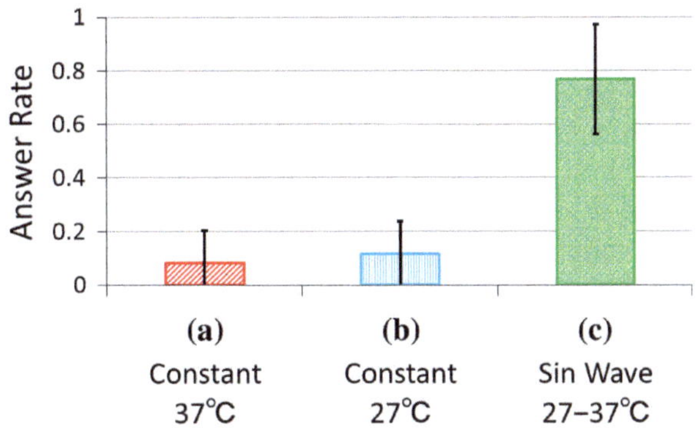

Fig. 2 Answer rate of vertical sensation occurrence

2.4 Result

Figure 2 shows answer rate of vertical sensation occurrence. In conditions (a) and (b) which presented constant temperature, occurrence rates were very low (around 10 %). In contrast, in condition (c) which presented constant temperature, occurrence rate was high (a little under 80 %). As result of Holm's law tests, there were significant differences ($p < 0.05$) between (a) and (c) and (b) and (c).

Interpretations of occurred sensation were various. The following were the answers for typical sensation.

- I take some to feel vertical sensation since touching stimulator.
- Vertical moving sensation elicited in synchronization with temperature change.
- Downward pulling sensation elicited when stimulus temperature declined.
- It was felt that the stimulator contacted by hand was expanding when stimulus temperature rose.
- Burning sensation was occurred when temperature changed.

All six participants' answers had in common that upward sensation elicited when stimulus temperature rose and downward sensation elicited when stimulus temperature declined.

3 Discussion

As a result of the experiment, it is certain that some kind of vertical sensation occurred. We focus that this illusion has similarities to active touch. When we touch some material, rapidly temperature change is occurred at the boundary between skin and material. Condition (c) which presented rapid temperature change recurred important factor of active touch. Therefore, participants may perceived tactile sensation like active touch when stimulus temperature was rapidly changing. There were associated studies that pressure sensation could elicit by high-temperature water vapor and controlling speed of temperature change when the moment of fingers touching some material could provide the illusion that like touching different material [2, 3].

In the experiment, it was reported that occurring burning sensation which was known as characteristic sensation of a kind of thermal illusion which is known thermal grill illusion (TGI) or synthetic heat (SH) [4–6]. TGI or SH was a well-known perceptual phenomenon that induces pain or burning sensations through the presentation of nonpainful hot and cold stimuli to the skin. Generally, TGI elicits simultaneous hot and cold temperature stimulation. However, TGI could occur by time differential two thermal stimuli [7]. Therefore, TGI might occur when stimulus temperature changed in the experiment.

4 Conclusion

We found that illusional force or tactile sensation occurred when temperature of thermal elements which were statically touched by subject's skin changes rapidly. In this paper, we conducted an experiment to verify incidence rate and quality of this thermal-tactile illusion. As result of the experiment, some kinds of vertical sensation were occurred and interpretations of occurred sensation were various (moving, tactile, or force). In the future, we compare between this thermal-tactile illusion and real force or movement.

References

1. ISO13732-1: Ergonomics of the Thermal Environment-Methods for the Assessment of Human Responses to Contact with Surfaces (Part 1: Hot Surfaces) (2006)
2. Kai, T., Kojima, Y., Hashimoto, Y., Kajimoto, H.: Mechanism of pressure sensation generated by hot steam. In: ISVRI (2011)
3. Yamamoto, A., Yamamoto, H., Cros, B., Hashimoto, H., and Higuchi, T.: Thermal tactile presentation based on prediction of contact temperature. J. Robot. Mechatron. **18**(3) (2006)
4. Green, B.G.: Synthetic heat at mild temperatures. Somatosens. Mot. Res. **19**(2), 130–138 (2002)
5. Craig, A.D., Bushnell, M.C.: The thermal grill illusion: unmasking the burn of cold pain. Science **265**, 252–255 (1994)
6. Bach, P., Becker, S., Kleinböhl, D., Hölzl, R.: The thermal grill illusion and what is painful about it. Neurosci. Lett. **505**(1), 31–35 (2011)
7. Watanabe, R., Saito, N., Mori, Y., Hachisu, T., Sato, M., Fukushima, S., and Kajimoto, H.: Evaluation of roller-type itch-relief device employing hot and cold alternating stimuli. In: ACM CHI extended abstracts (2013)

The Effect of Frequency Shifting on Audio–Tactile Conversion for Enriching Musical Experience

Ryuta Okazaki, Hidenori Kuribayashi and Hiroyuki Kajimoto

Abstract We have applied a frequency-shifting method, which was proposed previously in the literature for mixer manipulation, with the aim of generating vibration-based feedback to enrich the listener's musical experience. Experimental results showed that the proposed method significantly increased the listener's evaluation of sound consisting of high-frequency components, while a relatively poor evaluation was observed for sound containing low-frequency components.

Keywords Audio–tactile interaction · Audio–tactile conversion · Vibrotactile

1 Introduction

Recent studies have suggested that several cross-modal relationships exist between the tactile and auditory senses. Suzuki et al. [1] reported that tactile roughness perception can be modified by adding a task-irrelevant sound. Yau et al. [2] showed that the subjective auditory intensity is affected by simultaneous presentation of tactile stimuli. Each of the factors of audio–tactile interactions, including phase, synchrony, and frequency, has also been studied [3–6]. Physiological studies have

R. Okazaki (✉) · H. Kajimoto
The University of Electro-Communications, Chofu, Japan
e-mail: okazaki@kaji-lab.jp

H. Kajimoto
e-mail: kajimoto@kaji-lab.jp

R. Okazaki
JSPS, Tokyo, Japan

H. Kuribayashi
Nikon Corporation, Tokyo, Japan
e-mail: hidenori.kuribayashi@nikon.com

H. Kajimoto
Japan Science and Technology Agency, Tokyo, Japan

H. Kajimoto et al. (eds.), *Haptic Interaction*, Lecture Notes
in Electrical Engineering 277, DOI 10.1007/978-4-431-55690-9_9

reported that tactile and auditory sensations share a common neural mechanism [7]. Additionally, similarities between the tactile and auditory senses in a higher order region, e.g. the "consonance" between the two modalities, have also been investigated [8–10].

Recently, some audio–tactile conversion methods that focused on the cross-modal relationships were proposed with the aim of improving and enhancing the value of the content or the user experience (UX) using tactile stimulation. Karam et al. [11, 12] developed the Emoti-Chair, which focused on the spatial processing of the frequency at the auricle and then applied it to the tactile presentation. Birnbaum and Wanderley [13] proposed a "natural" tactile feedback method for electronic musical instruments based on the analysis of the vibration characteristics of real instruments. Lee and Choi [14] proposed an audio–tactile conversion method that focused on the roughness and the loudness. Lim et al. [15] proposed an audio–tactile conversion method that only deals with a specific frequency range to suit the preferences of the users.

In this paper, we propose an audio–tactile conversion method that focused on the gap in the perceivable frequency range between the auditory and tactile ranges. It is well known that there is a huge gap between the perceivable frequency ranges of the auditory (20 Hz–20 kHz) and tactile (0–1000 Hz) senses. As a result, the direct conversion of sound is likely to be unperceivable with an increase in frequency.

To address this issue, we tried to "compress" the frequency range of the audio signal to create a tactile vibration with a frequency that was shifted one or two octaves down. Previous studies suggested that a one-octave-shifted vibration created from an audio signal eased the user's distinction of musical instruments [16, 17]. We use a similar method. However, because the auditory frequency range is much wider than the tactile range, a one-octave shift might be insufficient for music consisting of higher frequency components. Our research questions are as follows:

1. To determine whether frequency-shifted vibrations enrich musical experiences.
2. To assess whether one- or two-octave shifts are effective, and when this occurs.

2 Conditions

To answer these questions, we prepared a control condition that provides an audio signal as tactile vibration (control) and four different conditions: (1) the tactile vibration is shifted one octave down against the original audio signal (1OT); (2) the tactile vibration is shifted two octaves down against the original audio signal (2OT); (3) combination of the two tactile tracks (1 + 2OT); and (4) combination of the original audio signal, the one-octave-shifted vibration and the two-octave-shifted vibration filtered by a 250 Hz, $Q = 0.0$ band-pass filter (1 + 2OT + BP). We used the band-pass filter to reduce very low- or high-frequency components. We found in a preliminary experiment that application of the two-octave shift sometimes produced very low-frequency components, which typically leads to an impression that the audio–tactile experience was "muffled". Similarly, very high-frequency components sometimes lead to a tingling feeling.

We used Hayaemon software (http://en.edolfzoku.com/hayaemon2/) to generate the octave-shifted tactile signal.

In the next chapter, the evaluations of the four specified conditions were measured in psychophysical experiments.

3 Experiment

Apparatus: The set-up comprised a computer with two audio channels. One audio channel powered the two sides of a set of high-quality headphones (QuietComfort, Bose Inc., USA), with strong active noise cancellation; the other channel was connected to an audio amplifier (RSDA202, Rasteme Systems Inc., Japan) driving a vibrotactile transducer (Haptuator mark 2, Tactile Labs, Canada). The transducer was firmly attached to a mobile device via a plastic cover (iPod Touch, 5th generation, Apple Inc., USA). The total weight of the vibrating device was 113 g (Fig. 1, left).

Participants: Five participants (three men and two women), aged between 22 and 43 years, were involved in the experiment. Each reported no auditory or tactile impairment.

Stimuli: To observe the effects of the proposed method for various frequency ranges, we prepared three different kinds of music. (A) The sound of a music box containing high-pitched tones (Akaneiro-op.x03). (B) Classical music containing mid-range sound (The Moldau, or Vltava). (C) Jazz music containing low-frequency tones (Sleepin' Maple Syrup Jazz). We used the first 15–20 s of each musical piece for the experiments.

The experimental stimulation consisted of an audio signal/tactile vibration pair. The total experimental stimulation included 15 pairs of audio and tactile stimuli. The average amplitudes of the signals were set to be equal. The auditory stimulation amplitude was set at about 55 dB SPL. The maximum acceleration with this setting was approximately 1.5 g.

Evaluation: To determine whether the evaluation of the content was altered by the combination of tactile and audio track signals, we carried out an evaluation using the VAS (visual analogue scale) with the following five questions. The questions used here were based on the questions used by Lee and Choi [14], but were partly modified: *Q*1: *Temporal harmony*—"Did the vibrations match with the sound temporally?"; *Q*2: *Frequency harmony*—"Did the frequency of the vibrations match with the sound?"; *Q*3: *Comfort*—"Did you feel comfortable with the vibrations, enabling you to enjoy the sound?"; *Q*4: *Preference*—"Did you enjoy the vibrations when presented with the sound?"; and *Q*5: *Fun*—"Did the vibrations make the sound fun?"

Procedure: While seated, each participant wore headphones and held the device in both hands. Each of the 15 pairs of auditory and tactile stimuli was presented

Fig. 1 *Left* The main components of the apparatus, consisting of headphones, an amplifier, and the vibrating device. *Right* Overview of the experiment

simultaneously. After experiencing these stimuli, the participants were instructed to answer the five questions using the VAS. No time limit was set for the task, but all tasks were completed within 60 s. All 15 pairs of stimuli were presented twice but at random, giving 30 tasks in total per participant (Fig. 1, right).

4 Results and Discussion

The experimental results are shown in Table 1. The lines in the table represent each condition, while the columns represent questions. The figures in the table indicate the average of the VAS scaled from 0 to 10. To verify the differences when compared to the control condition, Friedman's test and repeated Wilcoxon signed-rank tests were performed. The p values obtained were then corrected by the Holm–Bonferroni method at a significance level of 5 % to avoid multiplicity concerns.

*Q*1: *Temporal harmony*: In this question, the participants assessed the temporal matching of the sounds and vibrations that they experienced.

In the case of the sound of the music box, the evaluations of the 1OT, 2OT, 1 + 2OT, and 1 + 2OT + BP conditions were significantly higher when compared to that of the control condition ($p < 0.05$). It is believed that the evaluation of the control condition decreased because the main frequency contained in the original sound of the music box was much higher than the perceivable range of the tactile signal.

In contrast, in the cases of jazz and classical music, no significant differences were found when compared to the control condition.

*Q*2: *Frequency harmony*: Here, the perceived frequency harmony between the sound and the vibration was evaluated. The evaluations of the 1OT, 2OT, 1 + 2OT, and 1 + 2OT + BP conditions were significantly higher than that of the control condition for the sound of the music box ($p < 0.05$). In the classical music case, the only significant difference was observed between the 2OT condition and the control condition.

Table 1 Average results for Q1: temporal harmony, Q2: frequency harmony, Q3: comfort, Q4: preference, and Q5: fun

Q1	Control	1OT	2OT	1+2OT	1+2OT+BP
Music box	1.02 (1.30)	4.16 (2.93)	7.15 (1.83)	6.60 (2.70)	7.87 (1.87)
Classic	6.35 (1.90)	6.51 (2.35)	8.28 (1.05)	7.21 (1.20)	7.90 (1.84)
Jazz	8.56 (0.96)	7.81 (1.46)	7.59 (1.17)	8.09 (1.67)	7.97 (0.94)
Q2	**Control**	**1OT**	**2OT**	**1+2OT**	**1+2OT+BP**
Music box	1.14 (1.57)	4.24 (2.47)	4.15 (2.30)	5.13 (2.86)	6.34 (2.17)
Classic	4.71 (1.85)	6.10 (2.38)	7.87 (1.34)	6.32 (1.49)	6.32 (3.33)
Jazz	8.96 (0.71)	6.45 (1.96)	7.14 (1.28)	7.32 (1.57)	6.68 (2.38)
Q3	**Control**	**1OT**	**2OT**	**1+2OT**	**1+2OT+BP**
Music box	2.33 (2.41)	3.87 (2.90)	4.72 (2.00)	5.34 (2.46)	6.78 (1.35)
Classic	4.33 (2.35)	6.42 (2.34)	7.60 (1.44)	5.42 (2.88)	6.85 (2.54)
Jazz	8.31 (1.86)	6.74 (2.21)	5.57 (2.65)	6.87 (2.36)	5.71 (2.62)
Q4	**Control**	**1OT**	**2OT**	**1+2OT**	**1+2OT+BP**
Music box	1.73 (1.92)	3.73 (2.81)	4.44 (2.23)	5.62 (2.49)	5.81 (1.72)
Classic	4.24 (2.78)	6.21 (2.52)	8.09 (1.40)	5.57 (2.61)	6.26 (2.91)
Jazz	8.33 (1.60)	6.73 (1.89)	5.19 (3.29)	6.45 (3.13)	5.64 (2.90)
Q5	**Control**	**1OT**	**2OT**	**1+2OT**	**1+2OT+BP**
Music box	1.64 (1.92)	3.84 (2.78)	3.75 (2.74)	5.42 (2.77)	5.93 (1.89)
Classic	4.16 (2.84)	5.64 (2.45)	7.76 (1.69)	5.96 (2.04)	5.98 (2.71)
Jazz	7.96 (2.10)	6.69 (1.69)	6.10 (2.32)	6.39 (2.95)	5.87 (2.86)

Figures in *parentheses* represent standard deviations. Significantly higher and lower conditions compared to control condition are highlighted in *dark grey* and *light grey*, respectively.

However, in the jazz music case, the evaluations of all four conditions (1OT, 2OT, 1 + 2OT, and 1 + 2OT + BP) were significantly lower than that of the control condition ($p < 0.05$).

Q3: Comfort, Q4: Preference, and Q5: Fun: The evaluation tendencies for these criteria were quite similar. In the case of the music box sound, the evaluations of the 1 + 2OT and 1 + 2OT + BP conditions were significantly higher than that of the control condition at each criterion ($p < 0.05$). A significant difference between the 2OT condition and the control condition for jazz music was also observed in the evaluation of fun.

Discussions: From the experimental results, application of octave-shifted tactile vibrations to music will increase the evaluation of that music in terms of temporal matching, frequency harmony, comfort, preference, and fun, particularly when the original music comprises high-frequency components, e.g. the sound of a music box. In particular, in the 1 + 2OT and 1 + 2OT + BP cases, their evaluations were significantly increased for all criteria when compared to those of the control condition. It is therefore indicated that the proposed method will improve the quality of music that mainly consists of high-frequency components.

However, there was no significant difference between the evaluation of the proposed method and the control condition in the case of music that mainly comprises mid-range sounds, such as classical music. This may be because the mid-range music originally contains a range of frequency components that is perceptible with the tactile senses, regardless of the octave shifting.

In the case of jazz music, which mainly contains low-frequency components, the evaluation was partly reduced by octave shifting. There are two possible reasons for this. The first is the effect of the very low-frequency components generated by repeated octave shifting. Because jazz music originally consists of low-frequency components represented by the bassline, very low-frequency components (<10 Hz) were produced by performing the octave shifting once or twice. As a result, this very low-frequency component might suggest a muffled quality to the participants. In the 1 + 2OT + BP condition, the band-pass filter was applied to avoid this very low frequency; however, the participants still reported the muffled sensation in their introspection reports, which suggests that the band-pass filter was not sharp enough. The second possible reason is the frequency response characteristic of the transducer that we used, which exerts its full performance for vibrations at more than 60 Hz.

5 Conclusion

In this paper, we aimed to create a tactile vibration to increase the evaluation of music corresponding to that vibration, with the idea of compressing the frequency range by octave shifting that was previously used for tactile musical instrument distinction. Experimental results showed that proposed method significantly increased the evaluation of sounds that consisted of high-frequency components in terms of temporal harmony, frequency harmony, comfort, preference, and fun. However, relatively poor evaluations were obtained for sounds with low-frequency components.

References

1. Suzuki, Y., et al.: Selective effects of auditory stimuli on tactile roughness perception. Brain Res. **1242**, 87–94 (2008)
2. Yau, J.M., et al.: Separate mechanisms for audio-tactile pitch and loudness interactions. Front. Psychol. **1**, 160 (2010)
3. Yau, J.M., et al.: Temporal frequency channels are linked across audition and touch. Curr. Biol. **19**, 561–566 (2009)
4. Wilson, E.C., et al.: Integration of auditory and vibrotactile stimuli: effects of phase and stimulus-onset asynchrony. J. Acoust. Soc. Am. **126**(4), 1960–1974 (2009)
5. Wilson, E.C., Reed, C.M., Braida, L.D.: Integration of auditory and vibrotactile stimuli: effects of frequency. J. Acoust. Soc. Am. **127**(5), 3044–3059 (2010)

6. Wilson, E.C., et al.: Perceptual interactions in the loudness of combined auditory and vibrotactile stimuli. J. Acoust. Soc. Am. **127**(5), 3038–3043 (2010)
7. Kayser, C., et al.: Integration of touch and sound in auditory cortex. Neuron **48**(2), 373–384 (2005)
8. Altinsoy, E.M., et al.: Cross-modal frequency matching: sound and whole-body vibration. In: Proceeding of HAID2010. LNCS, vol. 6306, pp. 37–45. Springer, Berlin (2010)
9. Yoo, Y., et al.: Consonance perception of vibrotactile chords: a feasibility study. In: Proceedings of HAID2011. LNCS, vol. 6851, pp. 42–51. Springer, Berlin (2011)
10. Okazaki, R., et al.: Judged consonance of tactile and auditory frequencies. In: Proceedings of WHC2013, pp. 663–666 (2013)
11. Karam, M., et al.: Designing the model human cochlea: an ambient crossmodal audio-tactile display. IEEE Trans. Haptics **2**(3), 160–169 (2009)
12. Karam, M., et al.: The emoti-chair: an interactive tactile music exhibit. In: Proceedings of CHI2010, pp. 3069–3074. ACM, New York City (2010)
13. Birnbaum, D.M., Wanderley, M.M.: A systematic approach to musical vibrotactile feedback. In: Proceedings of ICMC2007, pp. 397–404 (2007)
14. Lee, J., Choi, S.: Real-time perception-level translation from audio signals to vibrotactile effects. In: Proceedings of CHI2013, pp. 2567–2576. ACM, New York City (2013)
15. Lim, J.M., et al.: An audio-haptic feedback for enhancing user experience in mobile devices. In: Proceedings of ICCE2013, pp. 49–50 (2013)
16. Merchel, S., et al.: Tactile Identification of non-percussive music instruments. In: Proceedings of Forum Acusticum 2011, pp. 1257–1261 (2011)
17. Merchel, S., et al.: Touch the sound: audio-driven tactile feedback for audio mixing applications. J. Audio Eng. Soc. **60**(1), 47–53 (2012)

Part II
Tactile Devices and Rendering

A Flexible PDMS-Based Multimodal Pulse and Temperature Display

Simon Gallo and Hannes Bleuler

Abstract Surgical robotics is developing rapidly. It has numerous advantages for the patient such as small scars, shorter operation time, and shorter recovery time. While teleoperated robotics also has many benefits for the surgeon such as tremor reduction, force scaling, and comfortable working position, it completely isolates the surgeon hands, his/her natural force, and tactile sensors, from the patient, reducing the intuitiveness of the interface and sometimes leading to accidents. In an effort to restitute these sensations to the surgeon, various haptic displays have been designed rendering forces and tactile sensations at the level of the master device. We study multimodal haptic feedback and its effect on the intuitiveness of the teleoperated surgical interfaces. We present here the design and evaluation of a multimodal tactile display that reproduces pulse-like and thermal feedback on the user's fingertip. In addition, the multimodal display is combined with a commercial force feedback device, and a virtual telepalpation task is implemented.

Keywords Multimodal · Tactile · Thermal · Pressure · Flexible

1 Introduction

One of the major drawbacks of minimally invasive surgery (MIS) is that it does not allow the surgeons to perform direct tactile exploration of internal tissues. By palpating the internal tissues, the surgeon is able to detect hidden arteries by perceiving their pulse and even tumors by detecting changes in stiffness and the temperature rise of the tissue [1]. Tactile displays providing pulse feedback to the surgeon have been implemented [2, 3], and most of them are based on pneumatic actuation of a polymer body with a thin membrane. The pneumatic pressure inflates the membrane thus creating the pulse. These displays are flexible, an important

S. Gallo (✉) · H. Bleuler
Laboratory of Robotic Systems (LSRO), EPFL, Lausanne, Switzerland
e-mail: simon.gallo@epfl.ch

H. Kajimoto et al. (eds.), *Haptic Interaction*, Lecture Notes
in Electrical Engineering 277, DOI 10.1007/978-4-431-55690-9_10

feature for tactile devices. Several thermal displays have been designed [4, 5], and they rely on Peltier elements to cool and warm the user's finger. Peltier elements require cooling (often water cooling) and are solid state, thus making the device rigid. The device presented in this paper is a combination of these two types of actuation methods. It uses hydraulic actuation to create the pulse but also thermal feedback by controlling the temperature of the water and retains the flexibility of the pulse devices. Here, the design and characterization of the display and its integration in a virtual environment to perform telepalpation tasks are presented.

2 Design and Characterization

This device is an improvement of [6]. It can be separated into two loops, the tank regulation loop and (2) the display feedback loop, as shown in Fig. 1a.

1. Tank regulation loop
   Two tanks store hot and cold water, respectively. The water from both tanks is circulated using submersible pumps through two copper heat sinks (heat exchangers). These heat exchangers are assembled on the top and bottom surface of the Peltier element. When powered, the Peltier element will cool the water circulating in one heat exchanger while warming the one in the other, thus cooling and, respectively, warming the water in the two tanks by forced convection. Since Peltier elements are asymmetric, they heat up more than they cool down, an additional aluminum heat sink and dedicated fan are added on top of the hot heat exchanger to evacuate the extra heat on the hot side of the Peltier.
2. Display feedback loop
   Two pumps inject water from the hot and cold tank into a PDMS display. The water then flows back to the tanks. By changing the proportions of hot and cold water, it is possible to achieve a precise and quick control of the temperature at the display. A solenoid valve is placed between the display and the tanks. By closing the valve, the hydraulic circuit is closed and the pumps build up the pressure in the circuit, thus creating the pulse feedback.

The display Fig. 1c consists in a main flexible PDMS body with a hollow chamber. It is fabricated using a molding process. A thin 100 μm membrane is attached to the main body using plasma oxygen bonding. When the pressure builds up in the display chamber, the membrane inflates. A pressure sensor is used to control the pulse. The pulse bandwidth goes up to 2 Hz [6] and the amplitude up to 4.2 mm with an input pressure of 40 kPa and is linear with pressure. As the membrane is very thin, the temperature drops from the water to the user finger is negligible. A T-type thermocouple is inserted in the display to control the thermal loop. The display is able to achieve rapid temperature rise and fall times up to 0.2 ± 0.01 and 0.27 ± 0.03 s, respectively, for an 11 °C step [6].

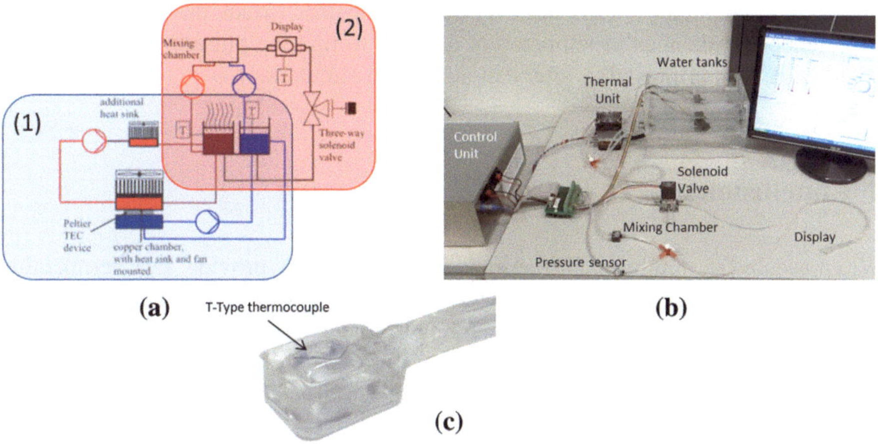

Fig. 1 **a** Schematics of the device function with the tank regulation loop (*1*) and the feedback loop (*2*), **b** picture of the setup and components, and **c** close-up picture of the flexible PDMS display

3 Telepalpation Task

For this palpation exercise, a virtual tissue with different temperatures and with hidden pulsations is created using CHAI3D (http://www.chai3d.org/). The virtual tissue illustrated in Fig. 2 is divided in three parts: the bottom left corner is a cold area without pulse feedback; the diagonal in the middle is at room temperature and a 1 Hz pulse feedback; and finally the top right corner has a high temperature with no pulse feedback. The PDMS display is strapped under the user index finger. The user then grabs the end effector of a commercial force feedback display, the Novint Falcon (http://www.novint.com/index.php/novintfalcon), thus experiencing force,

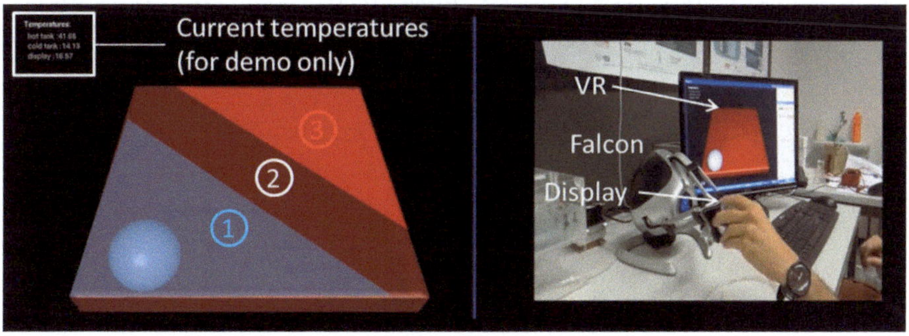

Fig. 2 *Left* Screenshot of virtual tissue with three different areas: cold (*1*), room temperature with pulse (*2*), and hot surface (*3*). *Right* Setup for telepalpation experiment

pulse, and temperature feedback. By navigating the virtual tissue, the user can experience the changes in temperature and has to find the pulsating region (which could be a hidden pulsating artery in a surgical simulation).

4 Conclusion

We presented a multimodal thermal display able to provide fast temperature changes and sensible pulses. This finger-sized display is PDMS based, and it is thus flexible and can be easily strapped on a finger, integrated in a glove or positioned at other locations on the body. The device was integrated in a manipulator operating on a virtual reality setup to perform telepalpation of a virtual tissue.

References

1. Guiatni, M., Riboulet, V., Duriez, C., Kheddar, A., Cotin, S.: A combined force and thermal feedback interface for minimally invasive procedures simulation. IEEE/ASME Trans. Mechatron. **18**, 1170–1181 (2013)
2. Santos-Carreras, L., Leuenberger, K., Rétornaz, P., Gassert, R., Bleuler, H.: Design and psychophysical evaluation of a tactile pulse display for teleoperated artery palpation. In: IEEE/RSJ International Conference on Intelligent Robots and Systems (IROS), pp. 5060–5066 (2010)
3. Streque, J., Talbi, A., Pernod, P., Preobrazhensky, V.: New magnetic microactuator design based on PDMS elastomer and MEMS technologies for tactile display. IEEE Trans. Haptics **3**, 88–97 (2010)
4. Gallo, S., Santos-Carreras, L., Rognini, G., Hara, M., Yamamoto, A., Higuchi, T.: Towards multimodal haptics for teleoperation: design of a tactile thermal display. In: 12th IEEE International Workshop on Advanced Motion Control (AMC), pp. 1–5 (2012)
5. Yamamoto, A., Cros, B., Hashimoto, H., Higuchi, T.: Control of thermal tactile display based on prediction of contact temperature. In: IEEE International Conference on Robotics and Automation. Proceedings. ICRA'04, pp. 1536–1541 (2004)
6. Gallo, S., Cucu, L., Thevenaz, N., Sengul, A., Bleuler, H.: Design and control of a novel thermo-tactile multimodal display. In: Haptics Symposium (HAPTICS), IEEE, pp. 75–81 (2014)

Adding Texture to Aerial Images Using Ultrasounds

Yasuaki Monnai, Keisuke Hasegawa, Masahiro Fujiwara, Kazuma Yoshino, Seki Inoue and Hiroyuki Shinoda

Abstract We present a method to add textures to aerial images using ultrasounds. The superposition of an acoustic radiation pressure on an aerial image allows users to feel a tactile texture of a virtual object floating in midair. Tactile textures can be altered by modulating the waveform of the ultrasounds. The tactile feedback is presented selectively on the user's fingertip using an infrared sensor without an extra marker. The proposed method provides a rich visuo–tactile experience all in free space.

Keywords Midair interaction · Noncontact tactile feedback · Floating touch panel · Aerial imaging

1 Introduction

Interactions with aerial images have suffered from the lack of tactile feedback [1]. Although other modalities such as audio–visual effects can be employed to indicate touch events, they do not act directly on the fingers [2]. Recently, we have proposed a system that superposes tactile feedback on aerial images using ultrasounds [3]. In this demonstration, we add rich tactile textures to the aerial images by modulating the ultrasound waveform. The system will provide a rich visuo-tactile experience all in free space.

Y. Monnai (✉) · K. Hasegawa · M. Fujiwara · K. Yoshino · S. Inoue · H. Shinoda
Graduate School of Frontier Sciences, The University of Tokyo, Tokyo, Japan
e-mail: Yasuaki_Monnai@ipc.i.u-tokyo.ac.jp
URL: http://www.hapis.k.u-tokyo.ac.jp/

H. Kajimoto et al. (eds.), *Haptic Interaction*, Lecture Notes
in Electrical Engineering 277, DOI 10.1007/978-4-431-55690-9_11

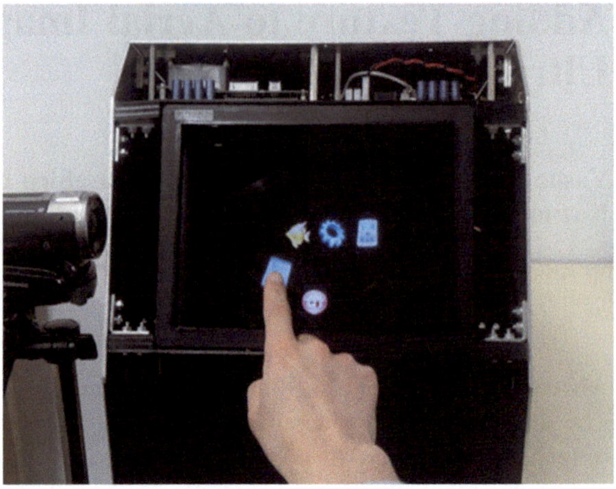

Fig. 1 System appearance

2 Setup

The system appearance is shown in Fig. 1. It consists of four components, an aerial imaging plate (AIP) [4], a liquid crystal display (LCD), an infrared (IR) touch sensor, and an ultrasonic phased array transducer. The floating virtual screen is generated by reflecting the LCD through the AIP. Finger insertion is detected with the IR sensor overlaying the virtual screen. Based on the acquired touch data, a focused ultrasound is delivered onto the fingertip by the ultrasonic phased array transducer. The phase and amplitude of each transducer is controlled independently for beam steering. In addition, the ultrasonic waveform can be modulated to present various tactile textures. For the ultrasound transmission, an indirect path reflected at the surface of the AIP is employed so that the ultrasound impinges on the virtual screen perpendicularly.

3 Application

In this demonstration, users are allowed to interact with aerial images, feeling significantly different tactile textures by modulating the waveform of the ultrasound. Each virtual icon, for example, can be equipped with different tactile textures for natural and precise input in midair. Rhythmical key input in midair is also enabled by physically indicating the finger stroke end. An application as a virtual drawing canvas allows users to draw graphics in midair while feeling the texture of the canvas.

References

1. Chan, L.W., Kao, H.S., Chen, M.Y., Lee, M.S., Hsu, J., Hung, Y.P.: Touching the void: direct-touch interaction for intangible displays. In: Proceedings of the ACM CHI, pp. 2625–2634 (2010)
2. Annett, M., Grossman, T., Wigdor, D., Fitzmaurice, G.: Medusa: a proximity-aware multi-touch tabletop. In: Proceedings of the ACM UIST, pp. 337–346 (2011)
3. Monnai, Y., Hasegawa, K., Fujiwara, M., Yoshino, K., Inoue, S., Shinoda, H.: HaptoMime: mid-air haptic interaction with a floating virtual screen. In: Proceedings of the ACM UIST (2014)
4. ASUKANET co. ltd. http://www.asukanet.co.jp

Driving System of Diminished Haptics: Transformation of Real-World Textures

Daisuke Yamaguchi, Yoichi Ochiai, Takayuki Hoshi, Jun Rekimoto and Masaya Takasaki

Abstract In this study, we developed a portable driving system for Diminished Haptics that transforms the haptic textures of real materials by ultrasonic vibration based on a squeeze film effect. A real material is attached on the display area (a square area with 30-mm sides), and its haptic texture is reduced. However, the vibration amplitude is decreased due to a shift of resonance frequency when user's finger contacts with the haptic display. Thus, a resonance frequency tracing system in the driving system plays an important role in maintaining the vibration amplitude. In this paper, we focus on the details of the resonance frequency tracing system and the effect of the system is evaluated.

Keywords Diminished haptics · Real-world texture · Texture transformation · Ultrasonic transducer driving · Resonance frequency tracing

---

D. Yamaguchi (✉) · M. Takasaki
Saitama University, 255 Shimo-Okubo, Sakura-ku, Saitama 338-8570, Japan
e-mail: yamaguchi14@mech.saitama-u.ac.jp

M. Takasaki
e-mail: masaya@mech.saitama-u.ac.jp

Y. Ochiai · J. Rekimoto
The University of Tokyo, 7-3-1 Hongo, Bunkyo-ku, Tokyo 113-0033, Japan
e-mail: wizard@slis.tsukuba.ac.jp

J. Rekimoto
e-mail: rekimoto@acm.org

Y. Ochiai
Japan Society for the Promotion of Science, 6 Ichiban-cho, Chiyoda-Ku, Tokyo 102-8471, Japan

T. Hoshi
Nagoya Institute of Technology, Gokisocho, Showa-ku, Nagoyashi Aichi 466-855, Japan
e-mail: star@nitech.ac.jp

J. Rekimoto
Sony CSL, 3-14-13 Higashigotanda, Shinagawa-ku, Tokyo 141-0022, Japan

H. Kajimoto et al. (eds.), *Haptic Interaction*, Lecture Notes in Electrical Engineering 277, DOI 10.1007/978-4-431-55690-9_12

1 Introduction

Representation of texture has attracted attention from manufacturer in industries, researcher of ergonomics, neuroscientist, and so on. Thus, the expression of textures has become a popular research area [1]. The aim of this study is to transform the textures of real objects. There are two major methods to transform real textures. One is to physically modify the original texture, and the other is to reduce or add the texture by actuation. In the real world, there are many types of haptic textures and it is possible to modify the textures by preprocessing, e.g., grinding and rapping. In this study, we focus on the second method for easiness and variability.

Previous methods for haptic texture representation are divided into two categories. One is wearable devices to provide additional vibration to users' fingers [2]. The other is haptic displays that provide haptic feedback on their smooth surfaces. The technologies employed in the latter approach include ultrasonic vibration [3, 4] and electrostatic forces [5], which have been applied to trackpads [6], pointing devices [7], and augmented reality (AR) systems [8].

In this study, we present a new method for haptic representation by reducing original textures. This method is opposite to previous methods. We reduce the degrees of real haptic textures based on a squeeze film effect of ultrasonic vibration [9]. This method is named "Diminished Haptics" [10].

The squeeze film effect is applied by using an ultrasonic transducer. Some real material is mounted on the tip of the transducer. The vibration amplitude is decreased due to a shift of resonance frequency when the material is contacted by user's finger. Thus, Diminished Haptics requires a function that traces the resonance frequency to keep the amplitude (i.e., thickness of a squeeze film) as much as possible. The driving system for this purpose is described in this paper.

2 Structure and Principle

The Diminished Haptics system consists of a vibration device (a transducer and a display surface) and a resonance frequency tracing system (a controller and a driver). The system overview is shown in Fig. 1 (Left). While the devices used in previous studies rendered some artificial textures, our device reduces the haptic textures on real objects. We use ultrasonic vibration to reduce the haptic textures based on the squeeze film effect. A haptic display surface on the vibration device is a square area with 30-mm sides. Real materials (e.g., sandpaper and wallpaper) are attached on the display surface. Users touch the display surface with their bare fingers and they feel reduction of the haptic texture by squeeze film effect.

Figure 1 (Right) shows the transducer (28 kHz resonance frequency without load) inside the vibration device, which is a bolt-clamped Langevin-type transducer with a horn. At a vibration node, the transducer is fixed on a casing with a jig.

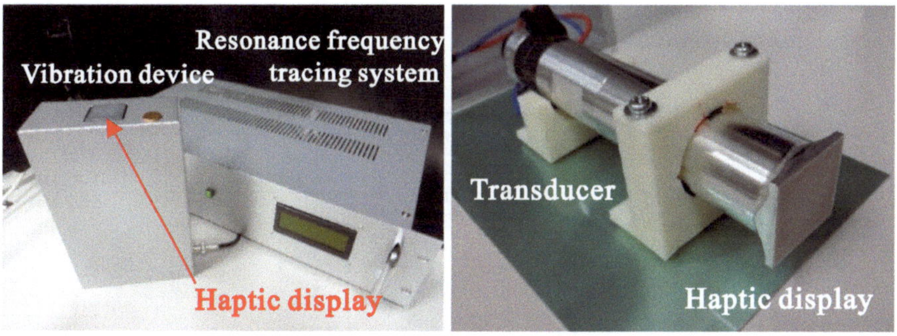

Fig. 1 *Left* Overview of diminished haptics system and *right* structure of vibration device

Users feel real textures without ultrasonic vibration. With ultrasonic vibration, thin air layers arise (squeeze film effect) and cover real materials. Users' fingers are slightly levitated, and the height of the real haptic texture is virtually reduced. The height of levitation can be controlled to continuously reduce the textures. The displayed textures have inherently high resolution and are felt without lateral movement of fingers differently from friction-control-based texture displays.

3 Resonance Frequency Tracing System

One of the issues of our method is that the resonance frequency of the bolt-clamped Langevin-type transducer is shifted when a finger contacts with the display surface. With a fixed-frequency driving signal, the vibration amplitude decreases due to this resonance frequency shift and cannot efficiently levitate fingers. Resonance frequency tracing is hence required to keep the amplitude.

This shift of the resonance frequency is evaluated by measuring frequency characteristics of admittance of the transducer in Fig. 2. These data were measured by an impedance analyzer. Due to the output limit of this analyzer, the measurement

Fig. 2 Admittance of transducer with/without finger

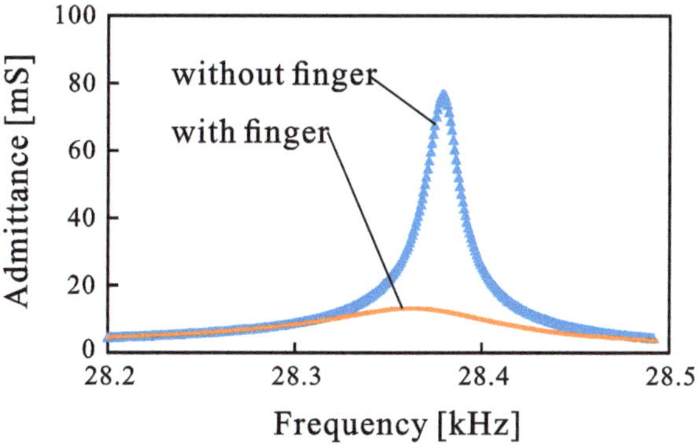

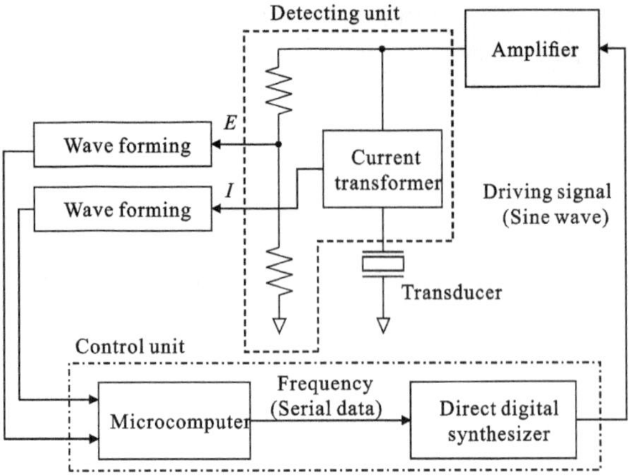

Fig. 3 Diagram of resonance frequency tracing system

was taken with a low voltage with which squeeze film effect is not arisen. While this condition is not equal to that in use, we can qualitatively evaluate the effect of finger contact. With finger, the shift and reduction of admittance are observed. Therefore, resonance frequency should be traced in order to keep effective the vibration amplitude. In this study, we developed a resonance frequency tracing system based on current phase measurement.

A diagram of the resonance frequency tracing system is illustrated in Fig. 3. This system consists of a microcomputer, a direct digital synthesizer, an amplifier, a voltage and current detecting unit, and two wave-forming circuits. These components are packaged in a portable casing. The phase difference at the resonance frequency is preliminarily measured and inputted to the tracing system as a target phase value. It traced the resonance frequency by monitoring the phase difference between the applied voltage and consumed current [11]. The detecting unit was inserted between the amplifier and the transducer to measure the phase difference. In the following experiments, this resonance frequency tracing system was used.

4 Experimental Result

To show the effectiveness of the resonance frequency tracing system, the vibration velocity of the display surface was measured by using laser Doppler vibrometer. The overview of the experiment and the measured velocity are shown in Fig. 4. The applied voltage was 19.1 $V_{0\text{-}p}$. Table 1 shows the measuredvibration velocities with tracing on/off and with/without a finger. The decrease of vibration velocity is observed with tracing off and with a finger. The reduction of real textures is not effectively arisen with this decreased vibration. On the other hand, the decrease of vibration with tracing on is negligible. These results indicate that our resonance frequency tracing system is effective in keeping the vibration velocity when a finger contacts with the display surface.

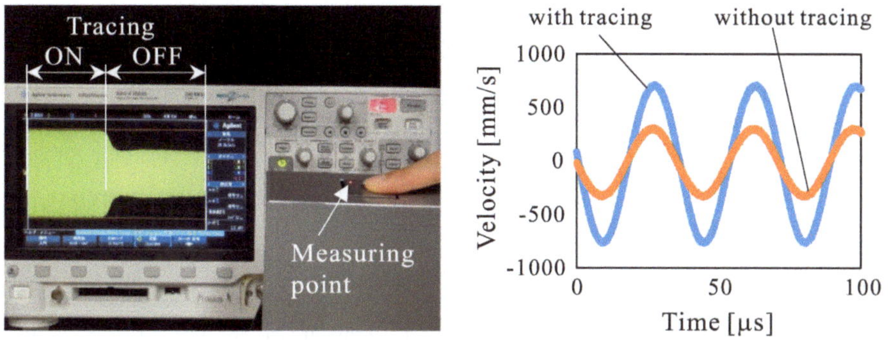

Fig. 4 *Left* Overview of experiment and *right* vibration velocity of display surface with and without resonance frequency tracing system

Table 1 Vibration velocity of tracing on/off and with/without finger

Tracing	Finger	Vibration velocity (mm/s)
OFF	Without	850
OFF	With	387
ON	With	818

5 Conclusion and Future Works

In this study, we fabricated and evaluated a portable driving system for Diminished Haptics. A transducer induces the squeeze film effect on a display area (a square with 30-mm sides). Real materials (e.g., sandpaper and wallpaper) are attached on the display area, and their haptic textures are reduced. To stabilize the degree of reduction of textures when a resonance frequency shift occurs due to placing a finger on the display surface, the system has a resonance frequency tracing system. We measured the vibration velocity to evaluate the effectiveness of the developed system, and we confirmed that it operated successfully. In the next stage, we plan to extend this method to 3D objects.

References

1. Hullin, M.B., Ihrke, I., Heidrich, W., Weyrich, T., Damberg, G., Fuchs, M.: State of the art in computational fabrication and display of material appearance. In: EUROGRAPHICS 2013, State of-the-Art Report (STAR) (2013)
2. Ando, H., Miki, T., Inami, M., Maeda, T.: Smart finger: nail-mounted tactile display. In: Proceedings of ACM SIGGRAPH, p. 78 (2002)
3. Biet, M., Casiez, G., Giraud, F., Lemaire-Semail, B.: Discrimination of virtual square gratings by dynamic touch on friction based tactile displays. In: Proceedings of 2008 Symposium on Haptic Interfaces for Virtual Environment and Teleoperator Systems, pp. 41–48 (2008)

4. Winfield, L., Glassmire, J., Colgate, J.E., Peshkin, M.: T-PaD: tactile pattern display through variable friction reduction. In: Proceedings of Second Joint EuroHaptics Conference and Symposium on Haptic Interfaces for Virtual Environment and Teleoperator Systems, pp. 421–426 (2007)
5. Bau, O., Poupyrev, I., Israr, A., Harrison, C.: TeslaTouch: electrovibration for touch surfaces. In: Proceedings of 23nd Annual ACM Symposium on User Interface Software and Technology, pp. 283–292 (2010)
6. Amberg, M., Giraud, F., Lemaire-Semail, B., Olivo, P., Casiez, G., Roussel, N.: STIMTAC, a tactile input device with programmable friction. Adjunct Proceedings. UIST'11, pp. 7–8 (2011)
7. Casiez, G., Roussel, N., Vanbelleghem, R., Giraud, F.: Surfpad: riding towards targets on a squeeze film effect. In: Proceedings of CHI'11, pp. 2491–2500 (2011)
8. Bau, O., Poupyrev, I., Goc, M.L., Galliot, L., Glisson, M.: REVEL: Tactile feedback technology for augmented reality. ACM Trans. Graphics **34**(1), 89–100 (2012)
9. Watanabe, T., Fukui, S.: A method for controlling tactile sensation of surface roughness using ultrasonic vibration. In: Proceedings of IEEE International Conference on Robotics and Automation, vol. 1, pp. 1134–1139 (1995)
10. Ochiai, Y., Hoshi, T., Rekimoto, J., Takasaki, M.: Diminished haptics: towards digital transformation of real world textures. In: Proceedings of Euro Haptics, no. 53 (2014)
11. Takasaki, M., Maruyama, Y., Mizuno, T.: Resonance frequency tracing system for Langevin type ultrasonic transducer. In: Proceedings of IEEE International Conference on Mechatronics and Automation, pp. 3817–3822 (2007)

High-Speed Thermal Display System that Synchronized with the Image Using Water Flow

Kyohei Hayakawa, Kazuki Imai, Ryo Honaga and Masamichi Sakaguchi

Abstract Man feels temperature in the cutaneous surface of not only fingers but also the whole body. There is a high possibility that temperature can be applied to information display. Temperature perception depends on the size of the area affected, the difference in temperature and the speed with which the temperature changes. In this study, we focused on the latter and developed a thermal comfort variable system. In this paper, we explain the characteristics of this system and the method of temperature presentation synchronized with the image.

Keywords Thermal display · Temperature control · Water flow · Haptic interface · Virtual reality

1 Introduction

The skin sensation has been actively applied to interface devices. However, many of them use force and/or pressure sensation. The temperature sensation is not used much. The reason is that heat flow is difficult to control, making temperature control difficult as well, for heat flow and temperature are directly related. So, it is difficult to realize a rapid change in temperature. For this reason, it takes a long time to transmit temperature information to a person and it is not perceived when temperature change is very slow.

K. Hayakawa (✉) · K. Imai · R. Honaga · M. Sakaguchi
Nagoya Institute of Technology, Nagoya, Aichi 466-8555, Japan
e-mail: cke13557@stn.nitech.ac.jp

K. Imai
e-mail: cju16515@stn.nitech.ac.jp

R. Honaga
e-mail: cix13145@stn.nitech.ac.jp

M. Sakaguchi
e-mail: saka@nitech.ac.jp

H. Kajimoto et al. (eds.), *Haptic Interaction*, Lecture Notes
in Electrical Engineering 277, DOI 10.1007/978-4-431-55690-9_13

Therefore, we can suppose that studies about temperature are expected in the future. Making that research into products/technology that can be used by customers/users requires an interface device that presents those changes in temperature.

Human's temperature perception depends on the size of the area affected, the difference in temperature and the speed with which the temperature changes [1]. Considering the use of that perception as an interface that allows the user to experience changes in temperature, there are limits to the size of the area and the temperature difference, since the human body has limited size and can only stand temperatures at a defined range. Therefore, for users to feel temperature changes clearly, it is necessary to realize temperature change in a short time. As mentioned above, since temperature change is accompanied by the movement of thermal energy, even though the temperature in the system can be changed immediately, the real change in temperature (hereby referred to as "response") does not occur fast enough.

The temperature of a skin surface can be quickly changed by touching objects that are heated or cooled beforehand. However, by this method, it will be accompanied by a contact sensation with an object. It is possible to present the temperature change without sense of touch by using a Peltier element.

A Peltier element is used in many cases [2], but a usual Peltier element has a speed of response that varies from several seconds to tens of seconds. In addition, it is necessary to process the waste heat.

In this study, we propose a high-speed temperature display system without contact sensation by using water as a heat medium [3]. Furthermore, we try to amplify the sense of reality by presenting temperature changes synchronized with an image. In this paper, we explain the characteristics of the temperature presentation system and the method of temperature change that is synchronized with the image.

2 Thermal Display System

2.1 Configuration

Figure 1 shows the developed thermal display system. Figure 2 shows a model diagram of the thermal display system. In this system, hot/cold water is sent to the presentation unit (D) by using 2 pumps (P_1, P_2) from the 2 tanks. The pump used was the KP-501T made by Koshin LTD. In the figure, the pink part shows the hot (plaid) water and the blue (brace) part shows the cold water. The switch between hot and cold water is carried out by the opening and closing of the solenoid valve (V_1–V_4). The solenoid valve is made by CKD Corp. Figure 3 shows the presentation unit, which is installed in a chair. It is made of vinyl chloride resin. When the water is flowing, it expands to about 47 mm in width, 120 mm in length, and 25 mm in thickness. It can fit many different parts of the body, because it is flexible.

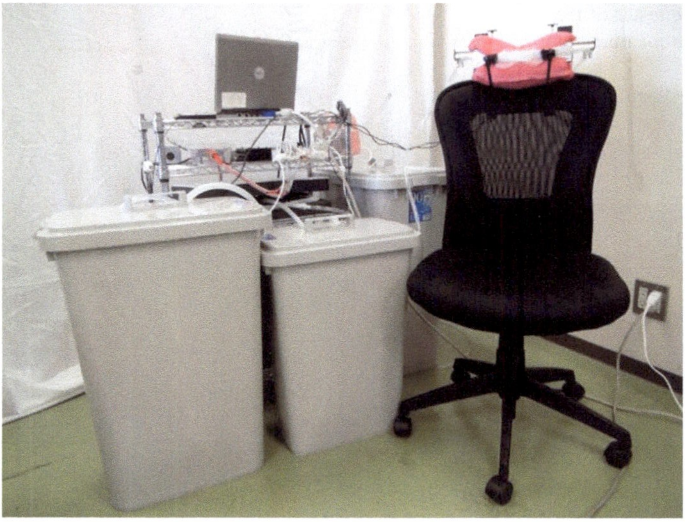

Fig. 1 Thermal display system

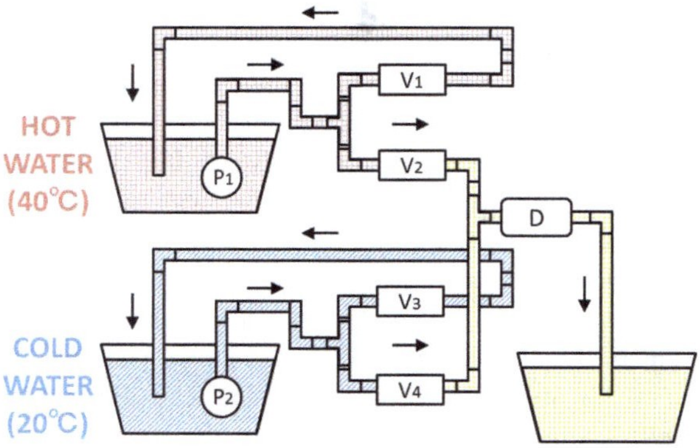

Fig. 2 Thermal display model

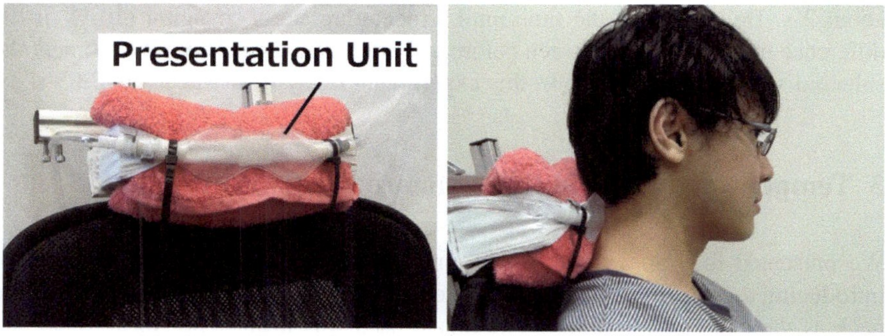

Fig. 3 Presentation unit

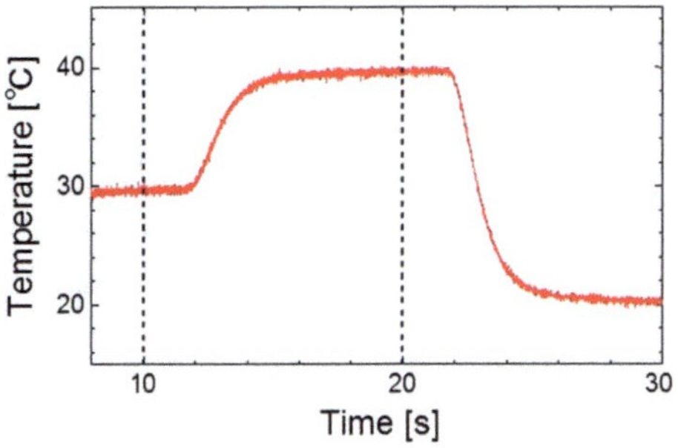

Fig. 4 The result of the temperature measurement

Table 1 Timing chart

Time (s)	Valve open	Temperature (°C)
0–10	V_2, V_4	30
10–20	V_2, V_3	40
20–30	V_1, V_4	20

2.2 Features

We use a thermocouple as a temperature sensor to measure the characteristics of the system. In this research, we used Type K thermocouple ST-55 made by RKC Instrument Inc. The diameter of the thermocouple line is 0.076 mm. The amplifier used was the AD595GQ made by Analog Devices Inc. It was connected to the computer through the A/D board, and temperature was measured.

Figure 4 shows the results of measuring the temperature change of the central portion of the presentation unit by using a thermocouple. In this experiment, we used 40 and 20 °C water, and hot and cold water valves were switched at 10-s intervals represented by dotted lines. Table 1 shows the timing of changes in temperature. It can be observed from Fig. 4 that the temperature has changed quickly. The rise/fall time was used as an evaluation index. The rise/fall time is defined as the time between 10 and 90 % of the difference between before and after a temperature change. As a result, the rise time is about 2.4 s and the fall time is about 2.2 s. Therefore, we can conclude that it is possible to change temperature in about 2 s. Delay time is the time until temperature starts changing (10 % of the difference in temperature between before and after the change) from the moment the solenoid valves are switched. In this experiment, delay time was about 2.05 s.

3 Temperature Presentation Synchronized with the Image

We presented the temperature synchronized with the image with the purpose of introducing a sense of reality. Figure 5 shows the experimental image. We chose the neck as the experiment surface. The image used was from "The North Wind and

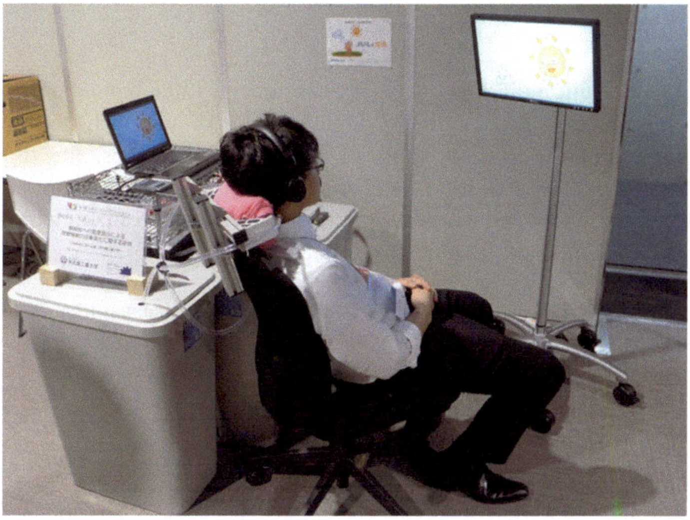

Fig. 5 Experience image

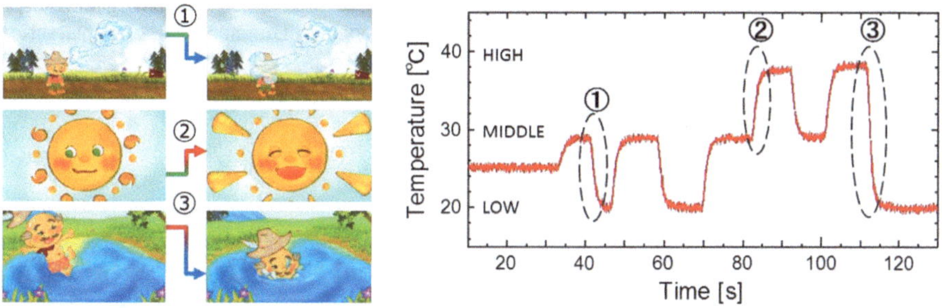

Fig. 6 Presentation scene of temperature change

The Sun." For example, as shown in Fig. 6, the temperature was lowered at the moment when the north wind was blowing and raised when the sun was shining.

So far, more than 150 people have experienced the temperature presentation that is synchronized with the image. We got positive comments from most of the participants. Some people perceived higher temperatures than the actual one when pumping hot water and lower temperatures with cold water. Some people got goose bumps when cold water was running.

4 Conclusion

In this study, we built a thermal display system using water as the heating medium. Since the presentation part of the system is made of vinyl chloride resin, it is possible to present the various spots of the body. Moreover, the difference in temperature of 20 °C was able to be changed in about 2 s. So far, more than 150 people have experienced the temperature presentation synchronized with the image. We received a lot of good comments.

Acknowledgments The image of "The North Wind and The Sun" that was used in this study was provided to MOONBEAMS CineArts & Science Inc. We would like to express our appreciation here.

References

1. Kanosue, K.: Encyclopedia of temperature and body. Asakura Publishing Co., Ltd., pp. 37–39 (2010) (in Japanese)
2. Sato, K., Maeno, T.: Presentation of rapid temperature change using spatially divided hot and cold stimli. J. Robot Mechatron **25**(3), 497–505 (2013)
3. Sakaguchi, M., Imai, K., Hayakawa, K.: Development of high-speed thermal display using water flow. In: Proceedings of 16th International Conference on Human–Computer Interaction, vol. 12. LNCS, vol. 8521, HIMI Part I, pp. 233–240 (2014)

Texture Modulation of 3D Fabricated Object via Electrotactile Augmentation

Shunsuke Yoshimoto, Yoshihiro Kuroda, Masataka Imura and Osamu Oshiro

Abstract We present a texture modulation technique that employs electrotactile augmentation to alter roughness perception during the texture exploration of a 3D fabricated object. The system detects user's touch action and renders the electrical stimulus that induces the modulating nerve activities at the middle phalanx of a finger. The users perceive fused sensation of the virtual sensation from the artificial electrical stimulus and the real sensation from the exploration of the fabricated object. The system allows users to explore the surface of a fabricated object with their bare finger and feel various roughness of the surface according to the modulation gain.

Keywords Texture modulation · Rapid prototyping · Fabricated object · Electrotactile display

1 Introduction

Fabrication technology (3D-milling, 3D-printing, etc.) makes a digital object tangible. Because the users can feel their design in their hand immediately, the technology is useful for the rapid prototyping of a product. If the texture of the fabricated object could be modified in real time, users can investigate various textures of the object to choose the best one. As for the visual texture of the object,

S. Yoshimoto (✉) · M. Imura · O. Oshiro
Graduate School of Engineering Science, Osaka University, Toyonaka, Japan
e-mail: yoshimoto@bpe.es.osaka-u.ac.jp

M. Imura
e-mail: imura@bpe.es.osaka-u.ac.jp

O. Oshiro
e-mail: oshiro@bpe.es.osaka-u.ac.jp

Y. Kuroda
Cybermedia Center, Osaka University, Toyonaka, Japan

H. Kajimoto et al. (eds.), *Haptic Interaction*, Lecture Notes in Electrical Engineering 277, DOI 10.1007/978-4-431-55690-9_14

there are several approaches using light projection technology to transform the appearance. However, once the object is fabricated, there is no way to transform touch feeling of the object surface without re-fabrication. Therefore, we are aiming at transforming the tactile texture of a fabricated object without re-fabrication.

There are several researches to provide transformed haptic/tactile feeling. Jeon and Choi [4] described a technique for modulating the stiffness of soft materials by augmenting additional force based on tool–object interaction. Hachisu et al. [3] proposed a method of modulating material perception using vibratory subtraction and addition. Changing material stiffness using artificial stimuli (e.g., jamming control [2]) is another possible method of haptic modulation. Asano et al. [1] succeeded in altering the fine and macro-roughness of real materials by applying vibrations to them. Ochiai et al. [8] proposed texture transformation by using squeeze film effect. The previous studies using artificial mechanical stimuli have reported excellent results in terms of user perception. However, it is difficult to eliminate mechanical constraints of materials and a finger because of size, location, and actuation of the stimulator. Thus, tactile texture transformation of the real fabricated object has not been achieved.

In this research, we propose a tactile texture modulation system for rapid prototyping by using electrotactile stimuli [5, 6]. The innovative approach of this work is tactile augmentation at the bare fingertip without disturbing texture exploration of the fabricated object. In this paper, we introduce how to build the texture modulation system using the electrotactile augmentation technology.

2 Tactile Texture Modulation System

The proposed system consists of a fabricated object, a positional sensor, a computer, and the proposed electrotactile display as shown in Fig. 1. The system detects user's touch action and renders the electrical stimulus that induces the modulating nerve activities at the middle phalanx of a finger to present various texture perception. The users perceive fused sensation of the virtual sensation from the artificial electrical stimulus and the real sensation from the exploration of the fabricated object.

2.1 Touch Detection

The stimulus is presented during the texture exploration of the fabricated object with a bare finger. For the contact detection between the finger and the object, a boundary mesh of the object is used. If the finger position is inside the boundary mesh, the modulating electrical stimulus is presented. The 3D position of the finger is detected by using optical tracking system comprising of a retroreflective marker and two infrared cameras (FLEX V100R2, OptiTrack) with lenses of focal length

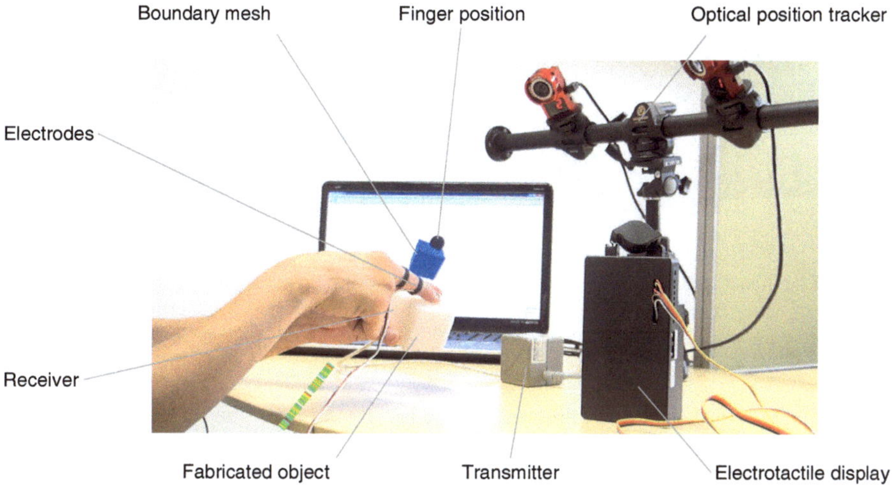

Fig. 1 The system architecture: modulating nerve activity is superimposed onto the afferent nerves at the middle phalanx of a finger by using anodic pulse stimuli

3.5 mm. The object's position and pose are detected by using a magnetic tracking system (3SPACE Fastrack, Polhemus). These two tracking systems are preliminary registrated to obtain the relative position. With this setting, the accuracy of the sensing is around 1 mm.

2.2 Electrotactile Display

In order to augment virtual sensation to the real sensation, we have developed an electrotactile display which can generate touch feeling at the fingertip without disturbing texture exploration [9]. As shown in Fig. 2, users wear the electrodes at their the middle phalanx of a finger to stimulate peripheral nerves. The principle is based on the fact that electrotactile sensation evoked by transcutanecous electrical stimulus [5, 6] is perceived at the site of the mechanoreceptors rather than that of the stimulated area of the nerves [7]. With a width of 200 μs and amplitude of up to 3 mA, the tactile display provides an electrical pulse stimuli density ranging from 1 to 100 pulses per second (pps). The pulse density and amplitude can be controlled by the computer via Bluetooth connection.

2.3 Stimulus Design

The electrical stimulus is controlled according to the velocity and pressure of the finger relative to 3D object. Specifically, we control the frequency of the

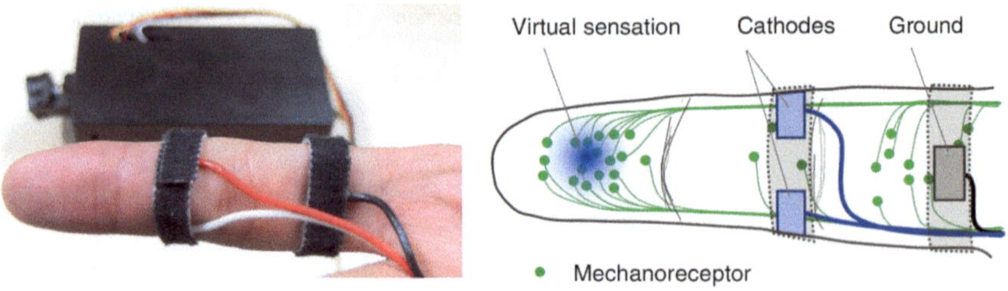

Fig. 2 Electrotactile display: two stimulus cathodes and one ground were located on the middle phalanx of a finger with a band. The users feel tactile sensation at the fingertip rather than the stimulated part

vibrotactile sensation by changing the pulse density of the electrical stimulus according to finger velocity as shown in Fig. 3. We also control the pulse amplitude according to finger pressure to equalize virtual and real sensation intensities. In practically, the finger pressure will be estimated by invasion amount of the finger position relative to the boundary mesh. Since the threshold of sensitivity varies between individuals, the pulse amplitude at each electrode has to be adjusted accordingly. Then, the calculated stimulus pattern is presented at each cathode.

2.4 User Experience

The system allows users to explore the surface of a fabricated object with their bare finger and feel altered roughness of the surface. The system can be applied for any fabricated object. In order to try various tactile textures, user can adjust the modulation gain. The modulation gain is a constant parameter of the velocity–density conversion. If the modulation gain becomes higher, the user would feel rougher sensation of the object as shown in Fig. 4.

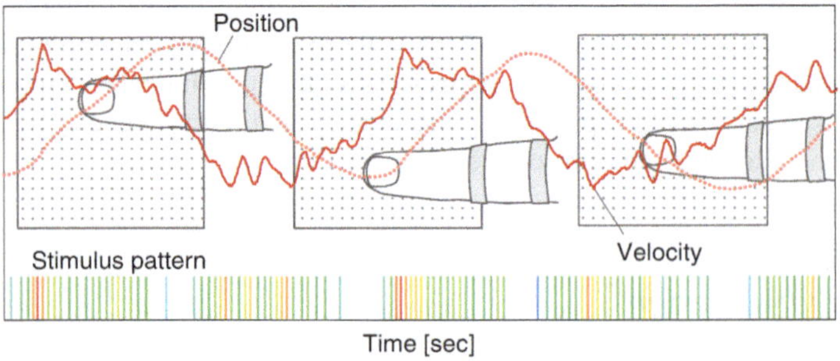

Fig. 3 Electrical stimulus pulse with finger position and velocity relative to the object: the pulse density of the electrical stimulus is controlled according to finger velocity

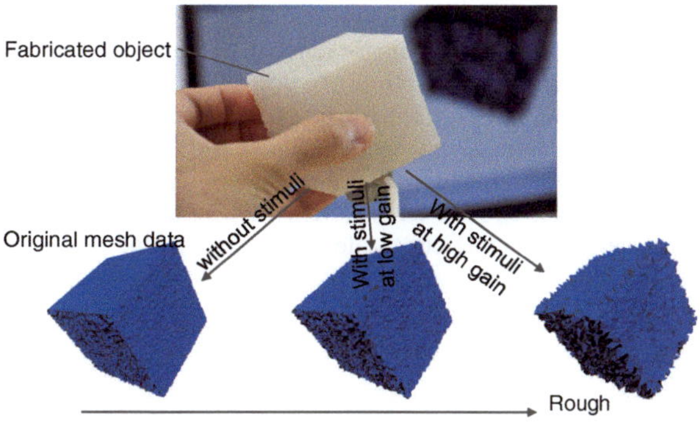

Fig. 4 Texture variation: the users can try various textures of the fabricated object according to the modulation gain

3 Conclusion

For the rapid texture prototyping, a tactile texture modulation system using electrotactile augmentation was proposed. Although it is known thatelectrical stimulus causes unpleasant sensation, the system alleviates the unpleasant sensation by combining with the real sensation from the exploration of the fabricated object. Since the proposed system enables users to feel various roughness of the fabricated object immediately, the system is expected to assist designing the texture of a product.

Acknowledgments This work was supported by JSPS KAKENHI Grant Numbers 22135003 and 25870394 in Japan.

References

1. Asano, S., Okamoto, S., Matsuura, Y., Nagano, H., Yamada, Y.: Vibrotactile display approach that modifies roughness sensations of real textures. In: Proceedings of IEEE RO-MAN, pp. 1001–1006 (2012)
2. Follmer, S., Leithinger, D., Olwal, A., Cheng, N., Ishii, H.: Jamming user interfaces: programmable particle stiffness and sensing for malleable and shape-changing devices. In: Proceedings of UIST, pp. 519–528 (2012)
3. Hachisu, T., Sato, M., Fukushima, S., Kajimoto, H.: Augmentation of material property by modulating vibration resulting from tapping. In: Proceedings of EuroHaptics, pp. 173–180 (2012)
4. Jeon, S., Choi, S.: Haptic augmented reality: taxonomy and an example of stiffness modulation. Presence Teleoperators Virtual Environ. **18**(5), 387–408 (2009)
5. Kaczmarek, K.: Electrotactile adaptation on the abdomen:preliminary results. IEEE Trans. Rehabil. Eng. **8**(4), 499–505 (2000)
6. Kajimoto, H., Kawakami, N., Tachi, S., Inami, M.: Smarttouch: electric skin to touch the untouchable. IEEE Comput. Graphics Appl. **24**(1), 36–43 (2004)

7. Micera, S., Keller, T., Lawrence, M., Morari, M., Popovic, D.: Wearable neural prostheses. Restoration of sensory-motor function by transcutaneous electrical stimulation. IEEE Eng. Med. Biol. Mag. **29**(3), 64–69 (2010)
8. Ochiai, Y., Hoshi, T., Rekimoto, J., Takasaki, M.: Diminished haptics: towards digital transformation of real world textures. In: Proceedings of the EuroHaptics conference 2014, p. 53 (2014)
9. Yoshimoto, S., Kuroda, Y., Uranishi, Y., Imura, M., Oshiro, O.: Roughness modulation of real materials using electrotactile augmentation. In: Proceedings of the EuroHaptics conference 2014, p. 15 (2014)

Rendering Different Sensations to Multiple Fingers in a Multi-digit Softness Display: Pulsation and Distributed Softness

Toshiki Kitazawa and Akio Yamamoto

Abstract This study demonstrates rendering of two unique softness sensations, by using a multi-digit softness display developed earlier by the authors. In one demonstration, a user can feel pulsating object within a soft material, which simulates a pulsating vessel or a heart within human body. The other demonstration is for distributed softness. By providing different softness sensations to multiple fingers, the system can render hard content within a soft body. In both of the demonstrations, a user can virtually rub the surface by providing lateral force to the device.

Keywords Softness display · Multiple fingers · Pulse · Lump · Distributed softness

1 Introduction

Human softness perception combines two different sensory information, which are kinesthetic and cutaneous. Recent studies revealed that cutaneous sensation is more dominant in softness perception. Therefore, if we intend to render softness sensations directly to bare fingers, cutaneous sense needs to be stimulated. So far, several softness displays have been developed to stimulate cutaneous sensations. In those displays, users can feel softness sensations directly on their fingertips, in contrast to conventional force displays in which users can feel softness only through styluses.

One of the most promising applications of such "bare-finger" softness displays would be regarding medical palpations. The softness displays can be employed as a palpation training tool, or as a local softness display in remote-palpation. In medical palpation, doctors use their fingers in many different ways. Sometimes, they use a single fingertip or multiple fingertips. For most of the cases, however, they use whole sections of multiple fingers. Looking at the reported softness displays, on the

T. Kitazawa (✉) · A. Yamamoto
The University of Tokyo, Tokyo, Japan
e-mail: kitazawa@aml.t.u-tokyo.ac.jp

H. Kajimoto et al. (eds.), *Haptic Interaction*, Lecture Notes
in Electrical Engineering 277, DOI 10.1007/978-4-431-55690-9_15

other hand, they typically limit the display area only to fingertips. For the above applications, such limitation does not seem satisfactory.

To allow a user to feel softness sensation on multiple fingers, Kitazawa et al. [1] have developed a multi-digit softness display. The device render softness to whole sections, not just fingertips, of three fingers, by wrapping flexible sheets around those three fingers. The device was designed such that the sheet wrapping for the three fingers can be independently performed, which means that it can render different sensations to different fingers. In this study and accompanied video, we demonstrate some applications using that function of the softness display. The demonstrations are rendering of pulse sensations and rendering of distributed softness.

2 Multi-digit Softness Display

This study utilizes the multi-digit softness display reported in [1], whose appearance is shown in Fig. 1. The device renders softness by changing contact area between fingers and the device, which is realized by wrapping flexible sheets around fingers. This rendering method was first proposed by Fujita and Ohmori [2] and was extended for wide variety of sensations by Kimura et al. [3] and Kimura and Yamamoto [4, 5]. To facilitate the sheet wrapping, the device is equipped with three dc motors, each of which controls wrapping height of the sheets for the corresponding finger.

3 Pulse Sensations

Pulse sensations can be sensed when we touch a specific part of human body, e.g., near a vessel or a heart. The device can render pulse sensations by adding a cyclic pulse wave to the wrapping height of the sheet. In the particular implementation of

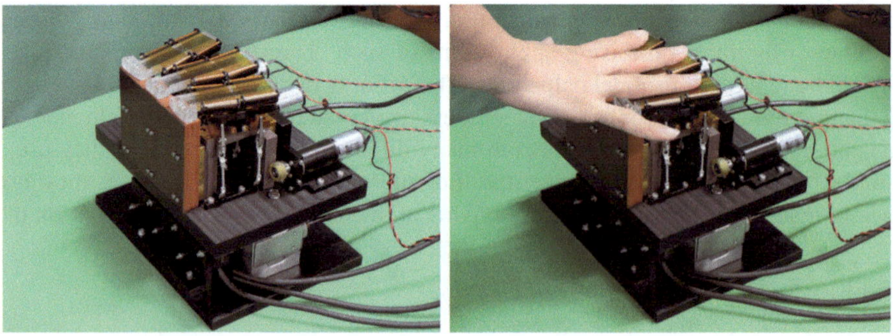

Fig. 1 Multi-digit softness display [1]

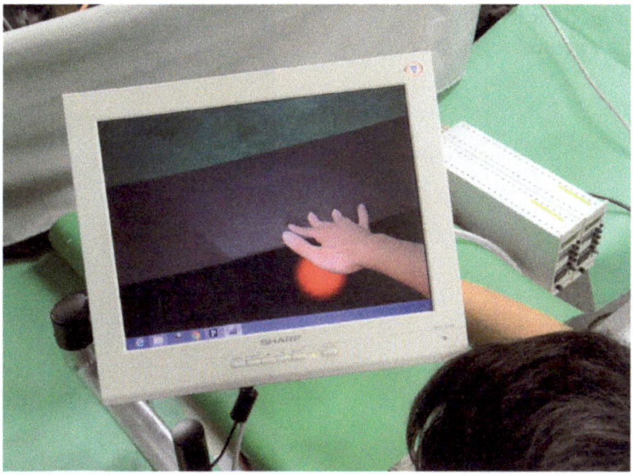

Fig. 2 Rendering of pulse sensations

this work, the softness display was combined with a visual display on which a photograph of a hand is displayed to enhance the reality, see Fig. 2.

The softness display is equipped with two load cells, which facilitate the measurement of the lateral force given from the user. The hand on a visual display is moved corresponding to the lateral force, which virtually allows the user to rub the surface of a soft object even though the position of the device is actually fixed. Within the soft object, a pulsating object is embedded. Based on relative locations of each finger against the pulsating object, the strength of the pulsation provided to each finger is determined; a finger closer to the pulsating object receives larger pulsation. In addition, the amplitude of the pulse wave was adjusted in proportional to the wrapping height of the sheet, such that a user can sense stronger pulse when she/he pushes stronger.

4 Distributed Softness

The other demonstration is rendering of distributed softness, which simulates, for example, a soft body containing a large lump. The implemented demonstration virtually allows a user to rub the lump from the surface of the soft body, as shown in Fig. 3.

The rendering is realized by controlling the sheet height to each finger, based on the thickness of the soft body (in this context, thickness means the distance from the surface of the soft body to the lump, at the location of each finger). To render a soft object whose thickness is finite, this work utilizes a contact theory proposed by Shull [6], in which they modified Hertzian contact theory considering the effect of a thickness of an object. By integrating the virtual rubbing described above, the system allows a user to feel the rough shape of a lump.

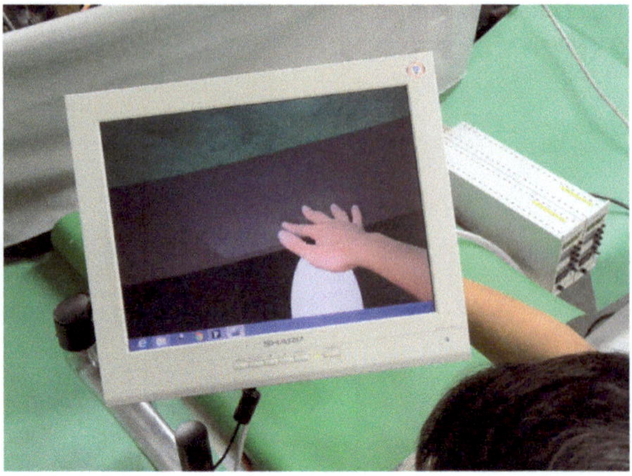

Fig. 3 Rendering of distributed softness

Acknowledgement This work was supported by Grant-in-Aid for Scientific Research (B) (No. 26280069) from JSPS, Japan.

References

1. Kitazawa, T., Kimura, F., Yamamoto, A.: Multi-digit softness: development of a tactile display to render softness feeling on multiple fingers. In: EuroHaptics 2014, (2014)
2. Fujita, K., Ohmori, H.: A new softness display interface by dynamic fingertip contact area control. In: 5th World Multiconference on Systemics, Cybernetics and Informatics (2001)
3. Kimura, F., Yamamoto, A., Higuchi, T.: Development of a 2-DOF softness feeling display for tactile tele-presentation of deformable surfaces. In: The 2010 IEEE International Conference on Robotics and Automation (ICRA). IEEE, New York (2010)
4. Kimura, F., Yamamoto, A.: Effect of delays in softness display using contact area control: rendering of surface viscoelasticity. Adv. Robot. **27**(7), 553–566 (2013)
5. Kimura, F., Yamamoto, A.: Rendering variable-sized lump sensations on a softness tactile display. In: World Haptics Conference (WHC), 2013. IEEE, New York (2013)
6. Shull, K.R.: Contact mechanics and the adhesion of soft solids. Mater. Sci. Eng. R Rep. **36**(1), 1–45 (2002)

Development of Wearable Outer-Covering Haptic Display Using Ball Effector for Hand Motion Guidance

Vibol Yem, Mai Otsuki and Hideaki Kuzuoka

Abstract An outer-covering haptic display (OCHD) is a device that imparts a guiding force sensation to the back of a learner's hand and guides the learner to manipulate a tool. Our previous study found that OCHD provides a skin deformation sensation and is able to guide a learner with less drive force than the alternative method where the tool is directly actuated. In this study, we developed a wearable outer-covering haptic display (wOCHD) for hand motion, with two ball effectors to deform the skin and provide a guiding information in four axes of motion.

Keywords Hand motion guidance · wOCHD · Ball effector · Skin deformation

1 Introduction

Laryngoscopy is one of the hand skills that require strength. It is a medical procedure in which a physician or a paramedic manipulates a tool called laryngoscope (Fig. 1, right) to open a patient's mouth to view the vocal folds and glottis.

Haptic systems are one of the major research areas to assist hand skills training to manipulate a tool. The most common method for such systems is to directly operate the tool using an actuator and a learner can directly sense the guiding force from the tool itself [1]. However, our previous study [2] found that this method is not suitable for hand skills training that requires the strong force for manipulation (i.e., when a learner needs to hold the laryngoscope tightly), because the guiding

V. Yem (✉) · M. Otsuki · H. Kuzuoka
The University of Tsukuba, 1-1-1 Tennodai, Tsukuba-shi, Ibaraki, Japan
e-mail: s1230202@u.tsukuba.ac.jp

M. Otsuki
e-mail: otsuki@emp.tsukuba.ac.jp

H. Kuzuoka
e-mail: kuzuoka@iit.tsukuba.ac.jp

H. Kajimoto et al. (eds.), *Haptic Interaction*, Lecture Notes
in Electrical Engineering 277, DOI 10.1007/978-4-431-55690-9_16

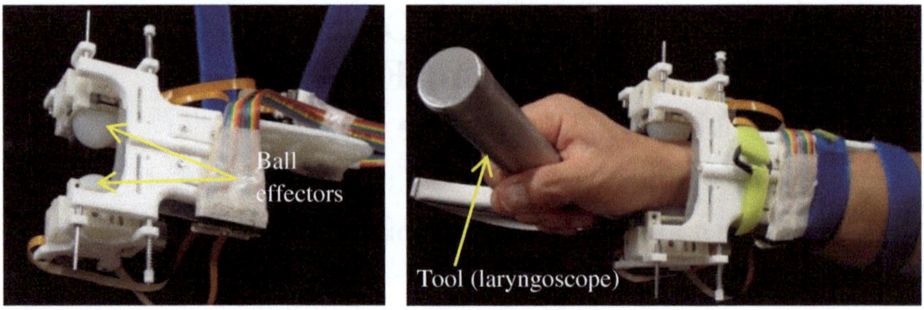

Fig. 1 wOCHD (*left*) and wOCHD worn on right hand (*right*)

forces inevitably become large, and this method typically requires a greater size and cost of the system.

Moreover, it is not an effective method for hand skills training because it is a passive learning technique, in which mainly the actuator manipulates the tool and the learner only follows its motion [3]. Although Saga et al. [4] proposed active learning method, in which learner tries to cancel the force produced by haptic device, in the case of laryngoscopy, it also requires a greater size and cost of the system.

To solve these issues, we proposed an outer-covering haptic display (OCHD), which imparts a guiding force to the back of a learner's hand by deforming the learner's skin [2]. Our previous studies showed that an OCHD is able to present learners in pitch rotation with less drive force than that is required for an actuator [2], and it is an effective method for hand skill learning [3].

The aim of this study was to develop a wearable OCHD (wOCHD) that uses ball effector to deform hand's skin. In this study, we used two ball effectors to provide guiding sensation in four axes of motion. Because wOCHD is a portable device, it does not require ground-based actuator to present guiding force to the learner's hand.

Regarding the wearable haptic device, Nakamura et al. introduced Hanger reflex device for presenting the pseudo-force sensation by skin deformation on wrist. The pseudo-force sensation makes user rotates his/her wrist in one axis [5]. On the other hand, we aimed to provide a guiding sensation in multiple axes of motion.

2 Wearable Outer-Covering Haptic Display

Kuniyasu et al. [6] proposed a wearable device that uses two plate effectors to deform the skin of the forearm and found that skin deformation can be used for training hand skills. Rotational skin stretch was also found to be effective for motion training of joint or limb [7]. In this research, we focused on applying skin deformation on the user's wrist using ball effectors (Fig. 1). The advantage of this method is that it can continuously deform the skin, even when slip occurs between the effectors and the skin.

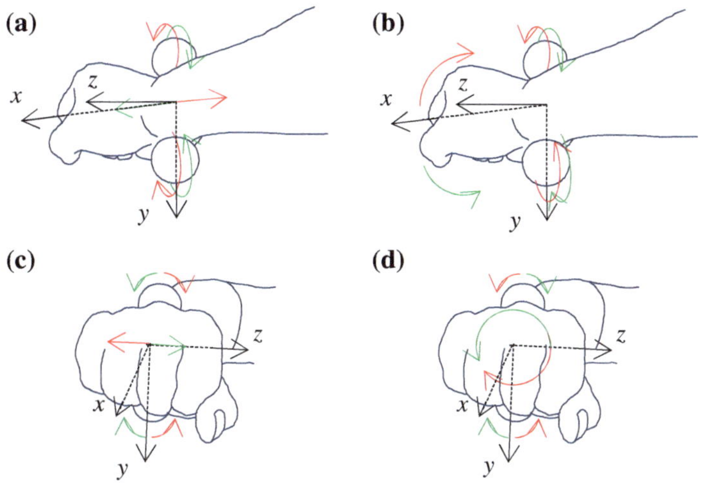

Fig. 2 Concept sketch of guidance in four axes (*x*, *yaw*, *z*, and *roll*). **a** Guiding in *x* axis. **b** Guiding in *yaw* rotation. **c** Guiding in *z* axis. **d** Guiding in *roll* rotation

Figure 2 shows how wOCHD controls two ball effectors to provide a guiding sensation in four axes of motion (*x*, *yaw*, *z*, and *roll*). As shown in Fig. 3, the actuator controls the ball effector in two degrees of freedom by using four shafts. Each pair of opposing shafts is moved by a single motor. Each ball was made from silicone and its diameter is 30 mm. Rotation angle of the ball was measured by hall sensors embedded in the motors. If there is no slippage between the ball and skin, the amount of skin deformation is equal to the amount of the rotation of the ball.

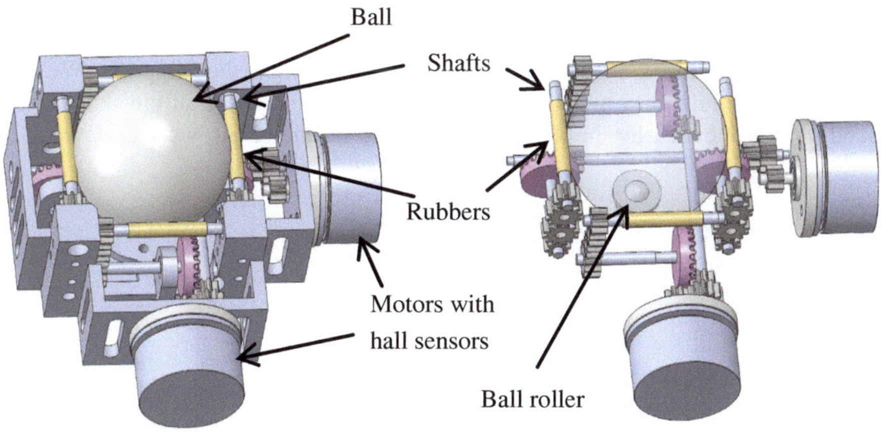

Fig. 3 Actuator (*left*) and four shafts' mechanism to actuate the ball (*right*)

3 Pilot Test and Discussion

For initial stage of our study, we conducted a pilot test to confirm that our wOCHD is able to provide a guiding sensation in four axes of motion. Four subjects participated in this test. Each subject was presented with skin deformation sensation in one of the eight random directions (positive and negative motion in each of the four axes). We assumed that there was no slippage occurred and the amount of skin deformation was about 10 mm. Each direction was presented three times, so that there were 24 trials in total for each subject.

The correct answer rates were 100 % for the *x* axis and 96 % for *yaw* rotation, but only 62 % for *roll* rotation and 60 % for *z* axis. The correct answer rate for *z* axis was low because they were confused with *roll* rotation, and vice versa. The subjects commented that the skin deformation strength of the two balls were not equivalent, which made it difficult to distinguish between the *x* axis and *yaw* rotation, and the *z* axis and *roll* rotation. We thought that this phenomenon occurred because of the different intensity of the sensation that depends on the location of the skin or the unequivalent pressure of the two balls. According to this result, in order to help subjects to perceive the guiding direction correctly, we need to add a force sensor to measure skin deformation more accurate even though the slippage occurred and a calibration to adjust the intensity of skin deformation sensation of the two balls.

4 Conclusion and Future Work

We developed wOCHD using two ball effectors to provide the skin deformation sensation to a user's wrist. The wOCHD is capable of guiding a hand in four axes of motion.

In our future work, we will improve our system by changing the effector's material to provide the equivalent pressure and reduce the slippage. We will investigate the relationship between the amount of deformation of the skin and the hand motion induced by the deformation. Based on the result, we will develop an

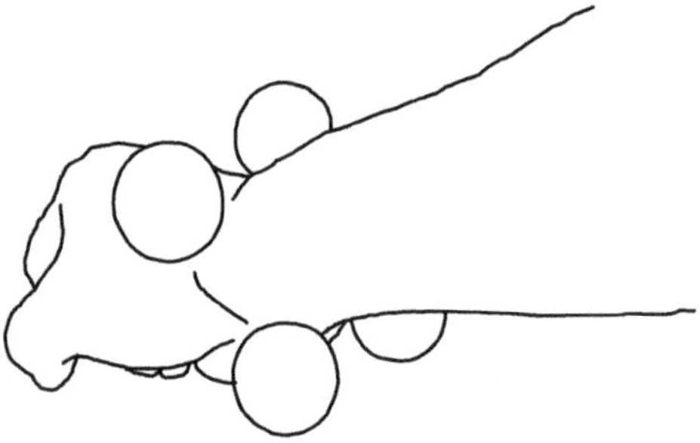

Fig. 4 wOCHD with six axes of motion using four ball effectors

algorithm for hand motion control. Additionally, in order to provide guidance with six axes of motion, we will develop wOCHD that has four ball effectors where additional two balls are for *y* axis and *pitch* rotation (Fig. 4).

Acknowledgement This work was supported by Grant-in-Aid for JSPS Fellows (25-2081).

References

1. Teo, C.L., Burdet, E., Lim, H.P.: A robotic teacher of Chinese handwriting. Proc. HAPTICS **2002**, 335–341 (2002)
2. Yem, V., Kuzuoka, H., Yamashita, N., et al.: Assisting hand skill transfer of tracheal intubation using outer-covering haptic display. In: Proceedings of ACM Conference on Human Factors in Computing Systems (CHI 2012), pp. 3177–3180 (2012)
3. Yem, V., Kuzuoka, H., Yamashita, N., Ohta, S., Takeuchi, Y.: Hand-skill learning using outer-covering haptic display. In: Proceedings of Eurohaptics2014, Article No. 50 (2014)
4. Saga, S., Kawakami, N., Tachi, S.: Haptic teaching using opposite presentation. In: Proceedings of World Haptic Conference (WHC 2005), CD-ROM, p. 1 (2005)
5. Nakamura, T., Nishimura, N., Sato, M., Kajimoto, H.: Application of hanger reflex to wrist and waist. In: Proceedings of IEEE Virtual Reality (VR 2014), pp. 181–182 (2014)
6. Kuniyasu, Y., Sato, M., Fukushima, S., Kajimoto, H.: Transmission of forearm motion by tangential deformation of the skin. In: Proceedings of Augmented Human International conference (AH 2012), Article No. 16 (2011)
7. Bark, K., Wheeler, J., Shull, P., et al.: Rotational skin stretch feedback: a wearable haptic display for motion. IEEE Trans. Haptics **3**(3), 166–176 (2010)

Presentation of Softness Using Film-Type Electro-Tactile Display and Pressure Distribution Measurement

Seiya Takei, Ryo Watanabe, Ryuta Okazaki, Taku Hachisu and Hiroyuki Kajimoto

Abstract Electro-tactile display has simple mechanical structure, and it can present tactile stimulus. However, electro-tactile display lacks feedback corresponding to touching motion. In this research, we developed a device that combines an electro-tactile display and pressure distribution sensor. The device not only solves the problem of unnatural electrical sensation, but also enables the expression of haptics-related physical characteristics, such as softness and viscosity. In an experiment, we validated whether the presentation of softness sensation is possible. As a result, presentation of softness sensation is confirmed that is possible by using pressure distribution measurement.

Keywords Electro-tactile display · Pressure distribution · Softness

S. Takei (✉) · R. Watanabe · R. Okazaki · T. Hachisu · H. Kajimoto
The University of Electro-Communications, 1-5-1 Chofugaoka, Chofu, Tokyo 182-8585, Japan
e-mail: takei@kaji-lab.jp

R. Watanabe
e-mail: r.watanabe@kaji-lab.jp

R. Okazaki
e-mail: okazaki@kaji-lab.jp

T. Hachisu
e-mail: hachisu@kaji-lab.jp

H. Kajimoto
e-mail: kajimoto@kaji-lab.jp

R. Watanabe · R. Okazaki · T. Hachisu
JSPS, Chiyoda-ku, Tokyo, Japan

H. Kajimoto
Japan Science and Technology Agency, Kawaguchi, Saitama, Japan

H. Kajimoto et al. (eds.), *Haptic Interaction*, Lecture Notes in Electrical Engineering 277, DOI 10.1007/978-4-431-55690-9_17

1 Introduction

Electro-tactile display is a tactile presentation device having simple mechanical structure. The interface part is just an electric substrate with electrodes, which can be as thin as 0.1 mm and flexible and can form curved shape [1, 2]. This greatly reduces limitation when combining with existing interfaces such as force display and visual display.

On the other hand, the electro-tactile display has several issues. One of those is lack of feedback corresponding to touching motion. In general mechanical tactile presentation, when the user presses the display stronger, the sensation becomes stronger. As we have learned this innate feedback in our daily lives, we feel it "natural." On the contrary, electro-tactile display does not have this innate feedback and typically, one feels the strongest sensation when he/she softly touches the electrode, due to insufficient contact. It leads to the general impression of "painful" or "frustrated" electrical sensation.

Electro-tactile display has the capacity to sense touch by resistance measurement for each electrode. However, in many cases, it can only detect touch, not the applied pressure. One previous work achieved pressure feedback by placing a force sensor under the electro-tactile display [3], but it could not measure the pressure of individual electrode. Therefore, presentation of tactile feedback in accordance with the fine pressure distribution was not possible.

To address this issue, we developed a device that combines an electro-tactile display and pressure distribution sensor. This device enables measurement of the pressure distribution applied by the users and use of the pressure distribution as input data for tactile feedback. Thus, the natural local feedback in accordance with pressure distribution becomes possible, which not only solves the problem of unnatural electrical sensation, but also enables the expression of haptics-related physical characteristics, such as softness and viscosity.

In this paper, after overviewing the system, we demonstrate that the pressure distribution measurement enables softness feeling, by presenting spreading sensation in accordance with pressure distribution change.

2 Experiment

In this experiment, we tested presentation of softness sensation by presenting tactile pattern in accordance with pressure distribution.

2.1 Device Structure

We use an electro-tactile display that has been developed in our previous work [2]. A film-type electro-tactile display unit with 192 (8 by 24) electrodes is driven by

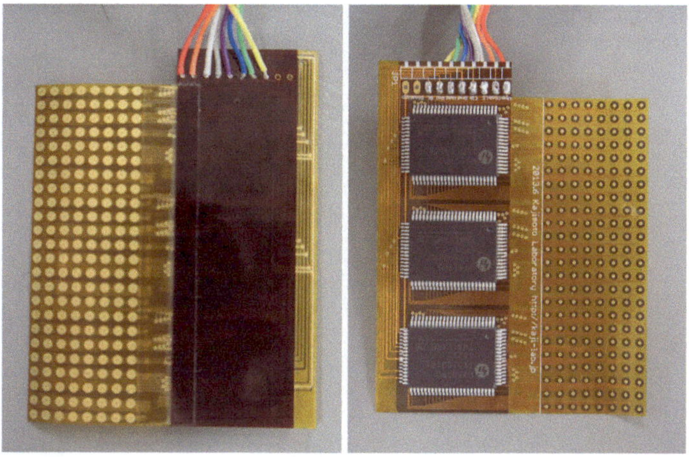

Fig. 1 Electro-tactile display

electrical pulse generator that communicates with PC (Fig. 1). Distance of each electrode is 3 mm. We also used a pressure distribution sensor unit that has 52 (4 by 13) measurement points (Fig. 2). Distance of each measurement points is 5.2 mm.

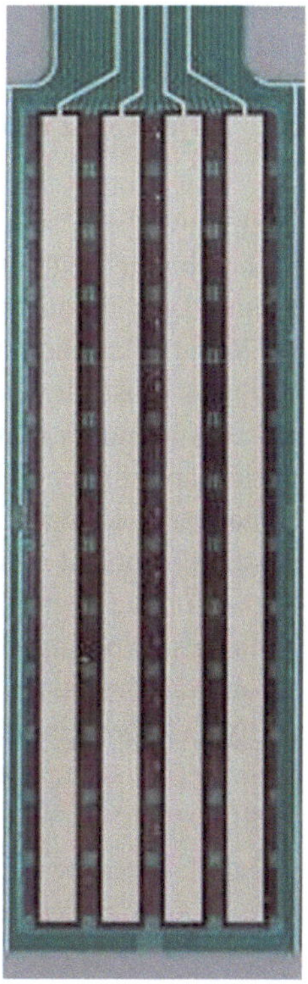

Fig. 2 Pressure distribution sensor

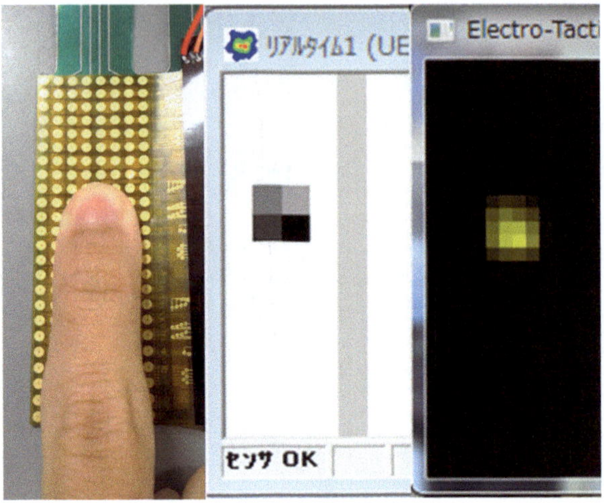

Fig. 3 State of pressure distribution measurement

The electro-tactile display is placed on the pressure distribution sensor, so that the system can simultaneously measure pressure distribution applied by the users and present tactile sensation. Figure 3 shows pressure distribution of the sensor (middle) and interpolated pressure distribution for each electrode (right). Pulse width was 30 μs, and frequency of measurement and stimulation loop was 50 Hz.

2.2 Softness Presentation Method

There have been several attempts to present softness, most of which paid attention to the change of contact area [4–6]. Soft objects deform rapidly so that the finger contact area becomes larger.

We also applied this principle to present softness by electro-tactile display. Figure 4 illustrates our methods. When the device is pressed strongly, the device presents stimulus that spreads like a ring to the outside of the contact area of the device and finger (Fig. 4, left, named Method 1). It enhances spreading of the contact area so that it should be perceived as soft. Figure 4 right (named Method 2) illustrates control condition, in which electrical stimulation is applied to the inside of the contact region.

In this experiment, we compared these two methods. In Method 1, the stimulation was presented at electrodes that are pressed with 1.35–9.021gf, which naturally achieve ring-shaped stimulation area. In Method 2, the stimulation was presented at electrodes that are pressed with the weight bigger than 9.021gf.

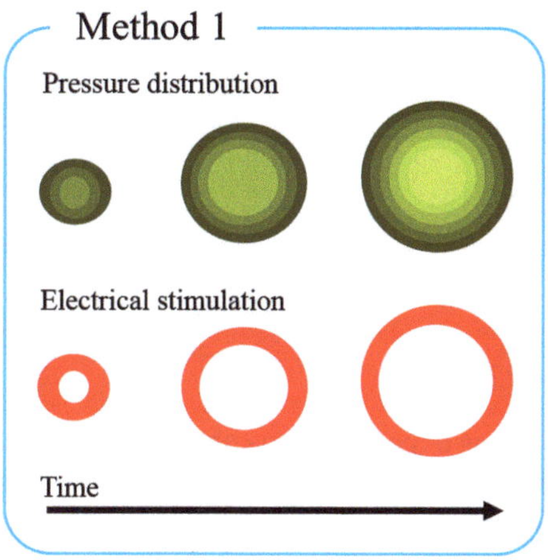

Fig. 4 Schematic view of stimulation methods

3 Procedure

We recruited five participants (all males, 22–25 years old) from our laboratory. First, participants pressed the device and adjusted the stimulus volume so that they do not feel pain. In subsequent procedure, the volume was fixed.

The main experiment was carried out by the following procedure.

1. Method 1 or Method 2 was presented to participants randomly, and the participants pressed the device by following visual indicator
2. The other method was selected, and the participants actively pressed the device by following visual indicator.
3. Participants chose which of the two was felt softer (forced choice)

The visual indicator is shown in Fig. 5. The top bar expands and contracts between 0 and 463.5 g, with 1 Hz sinusoidal curve. Participants controlled their pressure to match the bottom bar (participants' pressure) to the top bar. The experiment was conducted five times for one participant.

Fig. 5 Assistance of vision that presented to user

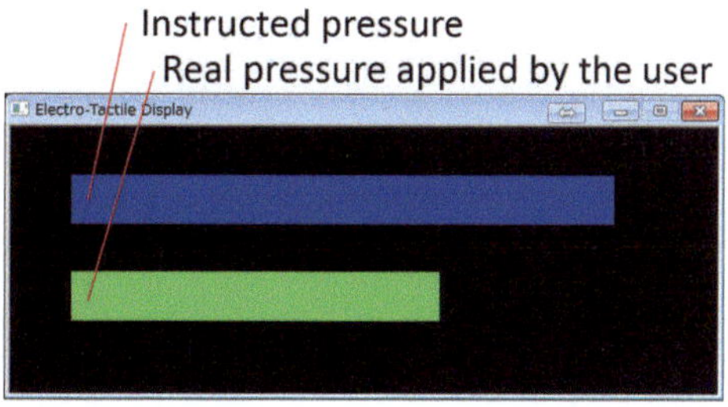

4 Result

Eighty-four percent of participants chose Method 1 as softer. We performed the paired comparison test of the ratio and found a significant difference between the chance level (50 %) and the result ($p = 0.001374$).

5 Conclusion

Based on the idea that electro-tactile display can express various physical properties by combining with pressure distribution information, we performed an experiment to present softness sensation using electro-tactile display and pressure distribution sensor. The result indicated that users felt softness sensation when electro-tactile display presented stimulus that spread like a ring when they pressed the electro-tactile display.

Our future work will include implementation of other physical properties such as viscosity and adhesion. We will implement the method to the application of interaction with virtual objects.

References

1. Kaczmarek, A.K.: The tongue display unit (TDU) for electrotactile spatiotemporal pattern presentation. Scientia Iranica **18**(6), 476–1485 (2011)
2. Kajimoto, H.: Design of cylindrical whole-hand haptic interface using electrocutaneous display. In: EuroHaptics, pp. 67–72 (2012)
3. Kajimoto, H., Kawakami, N., Maeda, T., Tachi, S.: Electro-tactile display with force feedback. In: World Multiconference on Systemics, Cybernetics and Informatics (2001)
4. Bicchi, A., Scilingo, P.E., De Rossi, D.: Haptic discrimination of softness in teleoperation the role of the contact area spread rate. IEEE Trans. Robot. Autom. **16**(5), 496–504 (2000)
5. Fujita, K., Ohmori, H.: A new softness display interface by dynamic fingertip contact area control. In: World Multiconference on Systemics, Cybernetics and Informatic (2001)
6. Kimura, F., Yamamoto, A., Higuchi, T.: Development of a contact width sensor for tactile tele-presentation of softness. In: Robot and Human Interactive Communication, pp. 34–39 (2009)

Development of Cold Sense Display Using Adjustment of Water Flow Volume

Ryo Honaga, Kazuki Imai, Kyohei Hayakawa and Masamichi Sakaguchi

Abstract In thermal display systems, the way to perform heating and cooling is an important issue. In this study, presenting temperature is controlled by adjusting the water flow volume in the thermal display system using water. In this paper, we explain the contents considering the water flow volume and the relation of the temperature changes, on the basis of the experimental result using the thermal displays of various form and size.

Keywords Thermal display · Water flow volume · Thermal control · Human interface

1 Introduction

Temperature is one of the environment parameters important for human. Therefore, many temperature controlling apparatuses, such as an air conditioner, an electric fan, a stove, and a pocket warmer, has been developed. But, their purpose is not to cause a temperature change but to keep temperature on steady values. Changes of temperature are used also for the guess of danger, and a possibility of being applicable to various kinds of information presentation is high.

R. Honaga · K. Imai · K. Hayakawa · M. Sakaguchi (✉)
Nagoya Institute of Technology, Nagoya, Aichi 466-8555, Japan
e-mail: saka@nitech.ac.jp
URL: http://ral.web.nitech.ac.jp/

K. Hayakawa
e-mail: 26413557@stn.nitech.ac.jp

H. Kajimoto et al. (eds.), *Haptic Interaction*, Lecture Notes
in Electrical Engineering 277, DOI 10.1007/978-4-431-55690-9_18

Human's thermesthesia is dependent on a difference in temperature, change speed, and presentation area. When using temperature as a human interface, the temperature region that can be used is limited because of viewpoint of safety. So, it becomes an important issue how a temperature change is shown correctly to use temperature as a human interface.

The purpose of thermal display system is controlling the temperature that a human body obtains and giving information from the sense of touch. They perform heating and cooling of thermal displays. Until now, researches that make a temperature change perception have been carried out. The research using Peltier device to make hot and cold space division stimulus and give high-speed temperature perception is one of them [1]. The Peltier device can show a temperature change without being accompanied by contact sensation with hot and cold objects. But, there is a problem about Peltier device in the viewpoint of limited presentation area. On the other hand, development of the high-speed thermal display system using hot water and cool water has also been performed [2, 3]. It has the merit that we can change the form of channel arbitrary and need not limit presentation area.

In this study, we control temperature, which is shown by adjusting water flow volume. We use water with low viscosity with large specific heat as a heat medium. And in this paper, we show the outline of the thermal display system which gives cold sense using water flow volume and the result of the temperature change by adjustment of water flow volume.

2 The Thermal Display System Using Water Flow Volume

The thermal display system that we made is a system that controls the temperature of the thermal display using water flow volume.

The reason we use water is that the flow can be adjusted easily thanks to low viscosity. Furthermore, its high specific heat makes its temperature change more easily when given heat from the human body.

2.1 The Whole Composition of System

Figure 1 shows the system-wide overview. The height of equipment is about 300 mm, the width is about 240 mm, and the depth is about 240 mm. Because the water jug with the cold-packed effect is used for the tank, the temperature of water in the tank is kept constant. The thermal display is stabilized on top, preventing that water in the tank escapes outside by putting on the tank.

Figure 2 shows the structure inside this system. The arrow in a figure shows the direction of movement of water. Water in the tank is sucked up with a bus pump KP-103 made from KOSHIN and it passes in the thermal display through a hose. Because of this, water flow volume in the thermal display changes and temperature

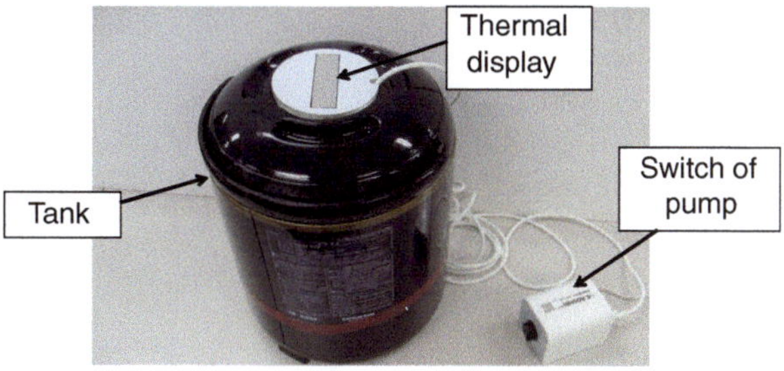

Fig. 1 Appearance of the system

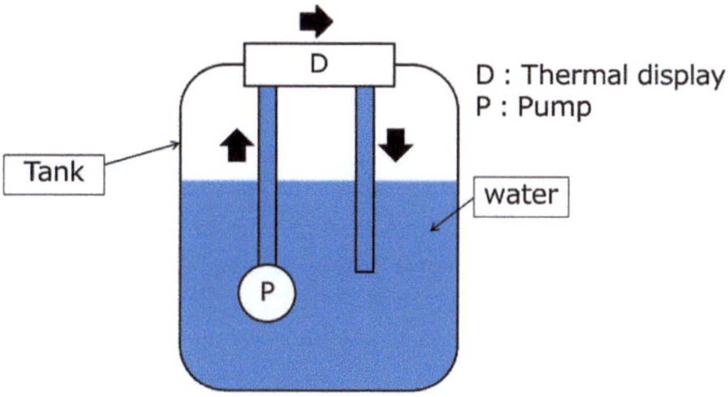

Fig. 2 The whole image of the system

changes are shown on the thermal display. The water that passes the thermal display flows into the tank through the hose again. Water repeats the circulation because the inside of the tank is closed space. By adjusting the drive of the pump, we can control the timing of water flows into the thermal display.

2.2 The Structure of Thermal Display

The structure of the thermal display consists of two acrylic boards and a aluminum sheet as shown in Fig. 3. The sheet of aluminum with high thermal conductivity is put on the channel cut in the acrylic board.

There are three variation thermal displays of 2 × 30, 3 × 20, and 5 × 12. In order to see the influence on temperature change about the depth of channel, we made three kinds of thermal display. We studied the influence of depth on temperature change using the thermal display of 3 mm depth as a reference. In addition, Table 1 shows the value in elements of the thermal displays.

The diameter of the thermal display is 94 mm, and the length of the channel is 55 mm. When the skin is touching to the surface of the aluminum board, temperature changes will be perceived at the time a stream occurs in the channel.

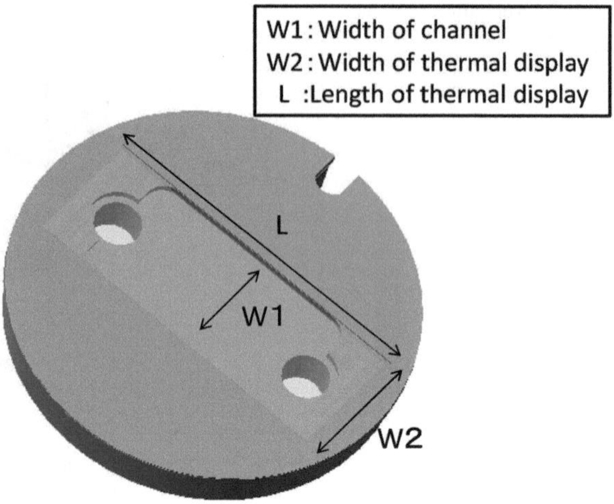

Fig. 3 CAD image (3 × 20 thermal display)

Table 1 Value in elements of thermal displays

	2 × 30	3 × 20	5 × 12
Width of channel (mm)	30	20	12
Depth of channel (mm)	2	3	5
Cross-sectional area of channel (mm^2)	60	60	60
Width of thermal display (mm)	38	28	20
Length of thermal display (mm)	77	77	75
Area of thermal display (mm^2)	2,916	2,156	1,500
Average of flow velocity (m/s)	1.68	2.08	2.01

3 Measurement of the Thermal Display's Surface Temperature

Figures 4, 5, and 6 each shows the results of surface temperatures measurement of the 2 × 30, 3 × 20, and 5 × 12 thermal displays, respectively.

In this measurement, we changed water flow volume with the fixed time cycle. We observed the temperature changes when carrying out the temperature presentation to human skin in this measurement.

Room temperature at the time of measurement was 25.8 °C, and water temperature was 25.0 °C. Moreover, surface temperature of the thermal display when human body touched was about from 30 to 31 °C. We used the thermocouple (ST-55) of type K made from RKC INSTRUMENT INC as a temperature sensor. In this measurement, we connected this thermocouple to the thermocouple amplification amplifier with cold junction compensation (AD595GQ). Furthermore, we connected the amplified voltage to the notebook PC through the A/D board. We measured temperature using the character of thermocouple that its output voltage has a high proportional system in temperature.

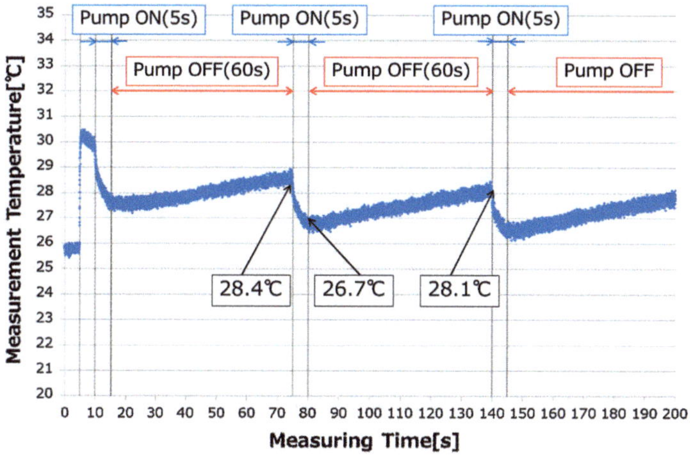

Fig. 4 Temperature on the 2 × 30 thermal display

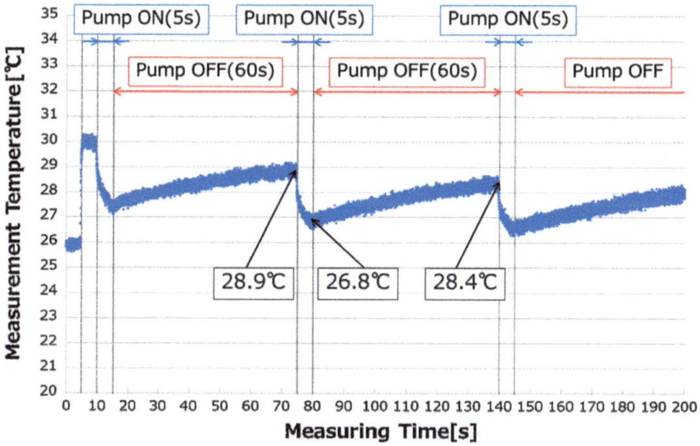

Fig. 5 Temperature on the 3 × 20 thermal display

In this experiment, at first 5 s, the thermal display was left on the state which air was touching. Next 5 s, it was changed into the state which skin was touching. After 10 s passed since the measurement start, the cycle that the pump's switch turned on for 5 s and turned off for 60 s was continued. In this cycle, the thermal display and the skin have been touching.

The large differences in temperature changes by the depth of channels were not seen in Figs. 4, 5, and 6. On the other hand, there was an opinion that the 2 × 30 thermal display felt like cooler than the 5 × 12 thermal display. In cross-sectional area regularity, the value of the width of the channel became larger if the depth of the channel became smaller. Therefore, the one which value of the depth of the

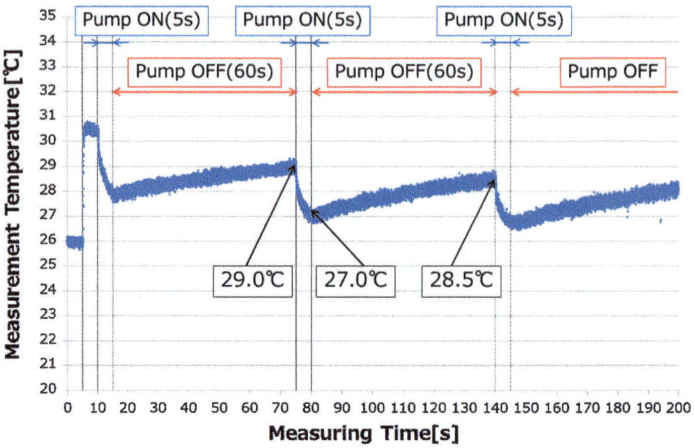

Fig. 6 Temperature on the 5 × 12 thermal display

channel is lower, the thermal display touches to wider range of skin. As a result, it becomes easy to perceive the temperature changes when the temperature changes are constant in all of the thermal displays. Therefore, the wide range of the channel is suitable form with the viewpoint of using as a human interface which shows temperature changes by adjusting water flow volume. Furthermore, we can select the form of thermal display by area we want to provide temperature change.

4 Conclusion

In this study, we constructed the thermal display system which causes a temperature changes by adjusting water flow volume. We constructed the various depth channels of thermal display on the basis of cross-sectional area regularity. After that, we measured and examined the surface temperature changes accompanying water flow volume changes. The large differences of changes in the surface temperature of the thermal displays about the depth of the channel were not seen. From this result, we can conclude that it is easier to perceive the temperature changes by making the channel thinner and wider. The reason is that the area of the thermal display touching to the skin becomes larger and, thus, the temperature change does not depend on channel's form.

From now on, we are going to utilize prevention of adaptation, to prevent drowsy, and to perceive risk function while driving by mounting on the car seat and cooling apparatus.

References

1. Sato, K., Maeno, T.: Presentation of rapid temperature change using spatially divided hot and cold stimuli. J. Robot. Mechatron. **25**(3), 497—505 (2013)
2. Imai, K., Hayakawa, K., Sakaguchi, M.: Development of the thermal comfort variable system by the temperature presentation to a neck. In: Proceedings of the 14th SICE System Integration Division Annual Conference, 3I1-2 (2013) (in Japanese)
3. Sakaguchi, M., Imai, K., Hayakawa, K.: Development of high-speed thermal display using water flow. Human interface and the management of information. Information and knowledge design and evaluation. Lect. Notes Comput. Sc. **8521**, 233–240 (2014)

HeatHapt Thermal Radiation-Based Haptic Display

Satoshi Saga

Abstract We propose HeatHapt, which is a concept of thermal radiation-based haptic display. By using thermal radiation and human characteristic of heat sensation, the HeatHapt realizes virtual haptic sensation in the air. The human feels drastic phase shift between 45 °C from hot to pain. We employ the characteristic and haptic movement of human and display haptic-like sensation to him. The system realizes haptic contact with digital contents directly with his hands.

Keywords Haptic display · Thermal radiation · Heat sensation · Nociception

1 Introduction

In recent years, spreading smartphones enhance anticipations toward haptic technologies. However, haptic interaction toward digital contents is very much limited. By employing thermal radiation and characteristic of human heat sensation, we focus on novel haptic display that realizes virtual haptic sensation in the air without any holdings.

Today many types of range sensors are available, such as Kinect, DepthSense, and Leap Motion. You can interact with many digital contents by your body. Several researches employ the device for virtual clay [1]. However, you cannot touch to the contents directly with your hands. There are no haptic feedbacks from the digital contents. Some researchers developed haptic display which can be used in the air [2]. However, the stimulation is not so strong.

S. Saga (✉)
Faculty of Engineering, Information and Systems, University of Tsukuba,
Tennodai 1-1-1, Tsukuba, Japan
e-mail: saga@saga-lab.org
URL: http://saga-lab.org/

H. Kajimoto et al. (eds.), *Haptic Interaction*, Lecture Notes
in Electrical Engineering 277, DOI 10.1007/978-4-431-55690-9_19

Fig. 1 Concept image of HeatHapt

On the other hand, interaction with deformable object, such as playing with clay, is very much essential for human creativity. Then, we realize such creativity with digital contents.

Here, we propose novel haptic display which enables us to contact with digital contents in the air without any gloves and holdings. We employ phase shift of heat sensation as virtual haptic contact (Fig. 1).

2 HeatHapt System

Here, we explain the working of the system (Fig. 2). The system consists of halogen lamp, pan/tilt mirror, thermal camera, and depth sensor. Halogen lamp emits powerful light to user's hand, and the hand is heated by the light. If the temperature of the hand exceeds 45 °C, the sensation of human changes from hot to pain (from heat sensation to nociceptive sensation). By employing the phase shift,

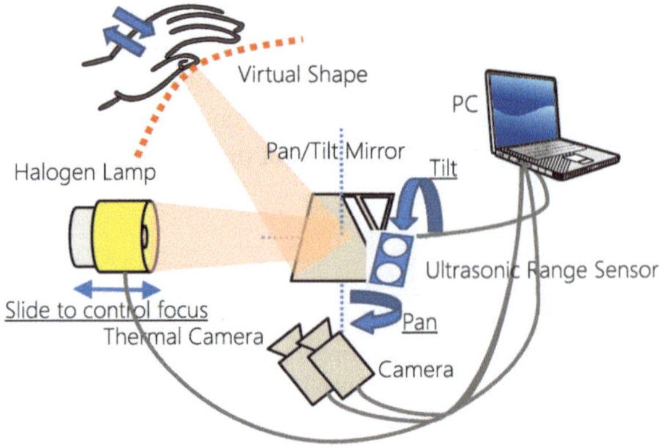

Fig. 2 Thermal radiation system

we convey haptic sensation to human. Pan/tilt mirror controls direction of emitted light, and on/off movement of reflection mirror controls the focus length of emitted light.

When the user put his hand in the air, the system measures the position of the hand. Normal camera and pan/tilt movement of ultrasonic range sensor detect three-dimensional position of the hand. The measured position and displaying virtual shape are compared in real time, and if there is no contact with virtual object, the pan/tilt mirror induces the thermal radiation to the hand. At the same time, based on the distance of the hand, the system slides the reflection mirror and controls the focus position of the radiation. In the same manner, if there is no contact with virtual object, the radiation to the hand is off. By using the thermal camera, we realize quick control between contact and no contact toward the object. The system controls the radiation to keep temperature on the hand around 45 °C, and it also avoids overheating of the hand. Thus, the user can feel nociceptive feedback on his hand according to his hand position. Around 45 °C, the sensation of human changes from heat sensation to nociceptive sensation. By employing the characteristic of human, we display haptic feeling in the air.

With the system, human can contact with digital contents directly through his hands, and in the future, he will exert enhanced creativity.

References

1. Han, Y.C., Han, B.J.: Virtual pottery: a virtual 3D audiovisual interface using natural hand motions. Multimed. Tools Appl. 1–17 (2013)
2. Hoshi, T., Iwamoto, T., Shinoda, H.: Non-contact tactile sensation synthesized by ultrasound transducers. In: EuroHaptics Conference, 2009 and Symposium on Haptic Interfaces for Virtual Environment and Teleoperator Systems. World Haptics 2009. Third Joint, pp. 256–260. IEEE, New York (2009)

Haptic Interaction on a Touch Surface

Dongbum Pyo, Semin Ryu, Seung-Chan Kim and Dong-Soo Kwon

Abstract This paper presents a new surface display that enables delicate tactile rendering with electrovibration and mechanical vibration. Haptic information on the surface can be expressed more richly if both feedback types are rendered simultaneously.

Keywords Surface haptics · Surface display · Electrovibration · Mechanical vibration · Haptic rendering

1 Introduction

Haptic interaction on surface-type human–computer interfaces, such as a touch screen, is growing more important. A touch screen with mechanical vibration feedback not only enhances fingertip interaction between the user and the interface, but also provides dynamic entertainment environments [1, 2]. Recently, numerous studies have been conducted in tactile feedback based on the electrovibration principle. Humans feel vibration or friction between a sliding finger and a surface through electrovibration [3–5]. However, because only fingers in sliding motion can receive this tactile feedback, users cannot be provided any feedback when they merely touch the surface. Mechanical vibration can play a complementary role for these types of tactile sensations with numerous patterns. Furthermore, more

D. Pyo · S. Ryu · S.-C. Kim · D.-S. Kwon (✉)
Department of Mechanical Engineering, KAIST, Daejeon, Republic of Korea
e-mail: kwonds@kaist.ac.kr

D. Pyo
e-mail: pyodb@robot.kaist.ac.kr

S. Ryu
e-mail: ryusm@robot.kaist.ac.kr

S.-C. Kim
e-mail: kimsc@robot.kaist.ac.kr

H. Kajimoto et al. (eds.), *Haptic Interaction*, Lecture Notes
in Electrical Engineering 277, DOI 10.1007/978-4-431-55690-9_20

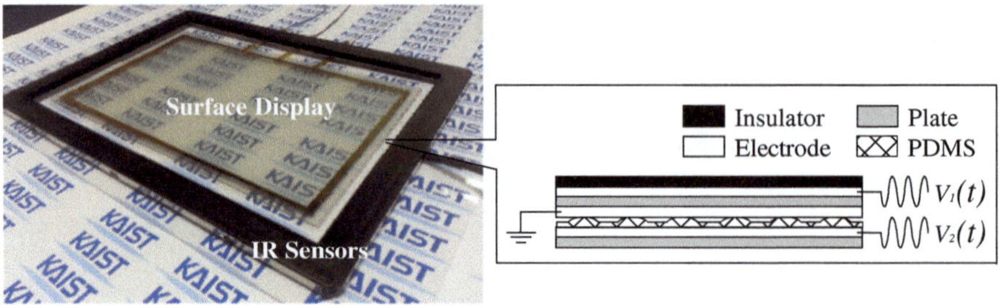

Fig. 1 Overall structure of the surface display

complicated and rich textures can be expressed delicately when electrovibration and mechanical vibration are rendered simultaneously. The proposed surface haptic display can provide effective tactile feedback from uniform mechanical vibration and electrovibration on an entire large surface.

2 System Implementation

Fig. 1 describes the overall structure of the surface display prototype. Electrovibration is generated on the surface of the top glass plate, which is coated with a transparent electrode and an insulating layer. Uniform mechanical vibration is produced by a periodic attractive electrostatic force between two parallel glass plates. The system consists of the surface display prototype, a control board as a signal source for electrovibration and mechanical vibration, high-voltage amplifier modules, an IR sensor module for sensing the fingertip position, and a desk PC, as shown in Fig. 2.

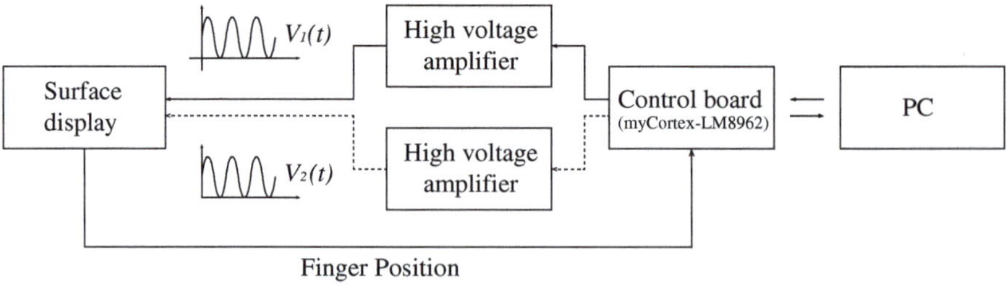

Fig. 2 Schematic representation of the system

3 Conclusion

In this paper, we introduced a new transparent surface display that can provide electrovibration and mechanical vibration simultaneously. Electrovibration combined with mechanical vibration increases the tactile information that can be expressed. We expect that the surface interaction concept using both types will enable users to experience abundant haptic effects in various applications.

Acknowledgements This work has been supported by the ICT R&D program of MSIP/IITP. [4-000-11-001, Human Friendly Devices (Skin Patch, Multimodal Surface) and Device Social Framework Technology].

References

1. Poupyrev, I., Maruyama, S., Rekimoto, J.: Ambient touch: designing tactile interfaces for handheld devices. In: Proceeding of UIST 2002, pp. 51–60. ACM, Paris, France (2002)
2. Jones, L.A., Sarter, N.B.: Tactile displays: guidance for their design and application. Hum. Fact. J. Hum. Fact. Ergon. Soc. **50**(1), 90–111 (2008)
3. Bau, O., Poupyrev, I., Israr, A., Harrison, C.: TeslaTouch: electrovibration for touch surfaces. In: Proceedings of UIST 2010, pp. 283–292. ACM, New York, USA (2010)
4. Meyer, D.J., Peshkin, M.A., Colgate, J.E.: Fingertip friction modulation due to electrostatic attraction. In: Proceedings of World Haptics 2013, pp. 43–48. IEEE, Daejeon, Korea (2013)
5. Kim, S.-C., Israr, A., Poupyrev, I.: Tactile rendering of 3D features on touch surfaces. In: Proceedings of UIST 2013, pp. 531–538. ACM, St. Andrews, United Kingdom (2013)

Friction Perception by Laterally Vibrotactile Stimulus: Early Demonstration

Akihiro Imaizumi, Shogo Okamoto and Yoji Yamada

Abstract Laterally asymmetric vibrotactile stimuli produce anisotropic or directison-dependent frictions and work as a friction display system. In this study, we showed that even symmetrical vibrotactile stimuli increased frictional perception based on the similar principle. We measured friction forces between a sliding finger pad and a contactor that was laterally vibrating at 3–5 Hz, and found that the lateral vibrations created zero relative velocity between the finger pad and the contactor and resulted in the frequent sticks between the two bodies. As a result, humans perceived stronger friction when scanning the vibrating contactor than scanning a stationary contactor.

Keywords Texture display · Skin stretch · Friction · Vibrotactile

1 Introduction

Recently, we have observed that perceived friction increased when scanning a contactor that asymmetrically vibrated in the lateral direction of a finger pad [3]. We measured the impulse of the frictional force applied to the finger pad sliding on the contactor, and it was larger when sliding in one direction than the other direction owing to the anisotropic frictional condition produced by asymmetric vibration. This anisotropy was also testified in a perceptual experiment. A possible principle of this phenomenon is that the finger pad more frequently sticks to the contactor in sliding one direction, and the sticking phase produces a friction force larger than the slipping phase.

By principle, even symmetric vibration such as a sinusoidal wave can cause the stick more frequently than a non-vibrating condition. In this study, we tested this

A. Imaizumi · S. Okamoto (✉) · Y. Yamada
Department of Mechanical Science and Engineering, The Graduate School of Engineering, Nagoya University, Nagoya, Japan
e-mail: okamoto-shogo@mech.nagoya-u.ac.jp

H. Kajimoto et al. (eds.), *Haptic Interaction*, Lecture Notes in Electrical Engineering 277, DOI 10.1007/978-4-431-55690-9_21

possibility in experiments. We measured the friction forces between the sliding finger pad and the contactor, when the contactor vibrated at 3 and 5 Hz, and it was still. By comparing the frictional forces among these conditions, we concluded that the symmetrically lateral vibrotactile stimuli were capable of causing the stick and slip phenomena, which led to the increase in the friction and its percepts. Although there are some earlier studies [1, 2, 4–6] about the relationships between the vibration and friction percepts, an attempt to increase the friction percept by a symmetric lateral vibration appears to remain to be studied.

2 Experimental Setup and Experiment

2.1 Lateral Vibrotactile Display

We established a vibrotactile display based on a DC motor (RE-40, Maxon) with an encoder (Type L, Maxon), as shown in Fig. 1. The contactor was made of finely polished ABS plastic that was fastened to the motor through a load cell (Model 1004, Tedea Huntleigh). We estimated a frictional force between the finger pad and contactor by subtracting the inertial force of the contactor from the force measured by the load cell.

2.2 Measurement of Frictional Force

We measured the friction forces while a finger slid on the contactor. Three conditions were tested, which were vibration at 3 and 5 Hz, and a still contactor. Each voluntary participant repeatedly scanned the contactor under these three conditions in one direction for a sustained amount of time. Figure 2 shows a sample of measured frictions. For comparison, the frictions for the still contactor (solid curve) and those

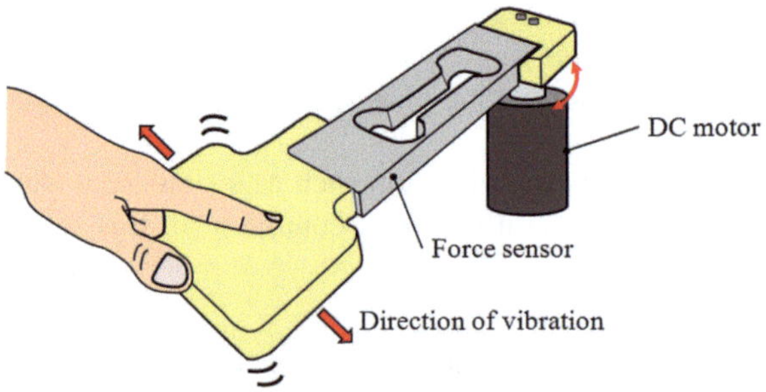

Fig. 1 Lateral vibrotactile display based on a DC motor. Equipped with a load cell. The finger scans the contactor along the direction of vibration

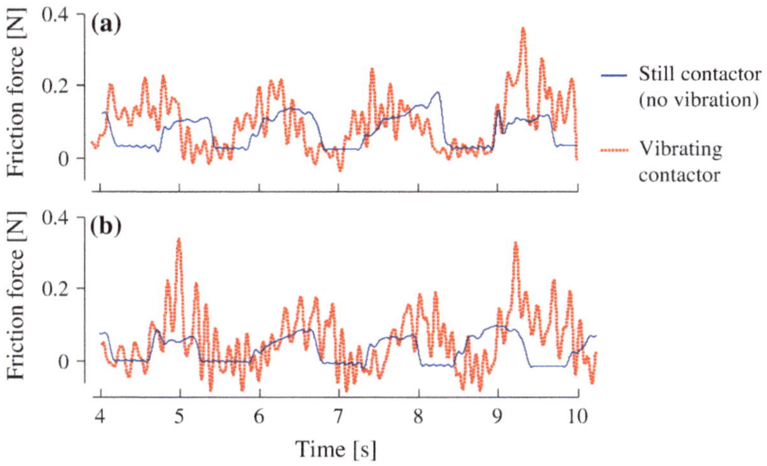

Fig. 2 Frictional forces when the finger scanned the contactor with and without vibrations of **a** 3 Hz and **b** 5 Hz. During the friction force being nearly zero, the finger was not in touch with the contactor. Sticks were more frequently observed between the vibrating contactor and finger than between the still contactor and finger. For visual clearance, the components larger than 10 Hz were filtered out

for the vibrating contactor (dotted curves) were plotted on the same figure. Note that, the figures also show the frictions while the finger was not in touch with the contactor. The periods with ignorable frictions were non-touched phases.

Frictions of the still contactor (solid curves) indicate that stick and slip phenomena hardly occurred when the remained contactor was slid by the finger because the fluctuations in frictions were minuscule. Some exceptions were observed at the beginning of sliding motions; however, the measured data corroborated that the kinetic friction was dominant for the still plastic contactor that we used in the experiment.

In contrast, the frictions (dotted curves) between the vibrating contactor and finger pad significantly fluctuated, which suggests that the sticking and slipping modes alternatively appeared. Since the maximum static friction is greater than the kinetic friction, the frictions observed from the vibrating contactor were largely greater than those from the remained contactor. Furthermore, such sticks between the finger pad and vibratory contactor were perceptually clear for all those who experienced the laterally vibrating stimuli.

3 Discussion and Conclusion

Larger friction forces were observed in scanning the vibratory contactor than in scanning the still contactor. Figure 3 shows a conjectured interpretation of this observation. In general, when the relative velocity between two bodies is not zero,

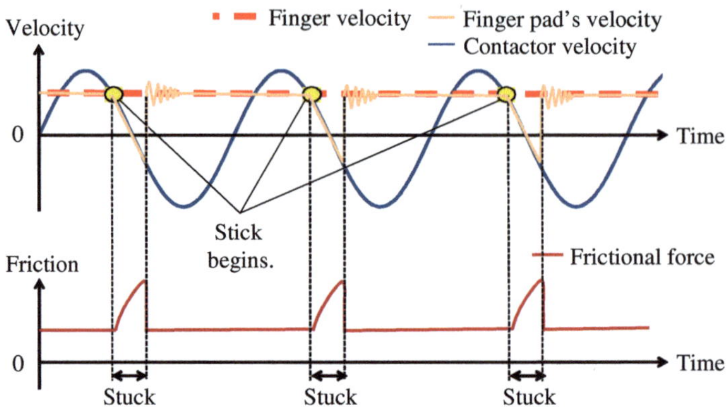

Fig. 3 Velocities of the finger and vibrating contactor and the frictional force between them

they are sliding. As described before, the slippage was considered to be predominant when the finger slid on the still contactor. On the other hand, when the finger slid on the vibrating contactor, the sign of the relative velocities periodically switched. This switch would have led to the stick of the two bodies and caused larger frictions. The friction increased synchronizing with the occurrence of sticks or vibration because the maximum static friction is greater than the kinetic one. The observed friction forces in Fig. 2 showed that the frequencies of peaky frictions are consistent with the vibratory frequencies of the contactor, which supports our interpretation.

In summary, it is speculated that when a finger slides on the laterally vibrating contactor, sticks occur between the two bodies and the friction percepts are increased. Understanding the underlying principles will be pursued.

Acknowledgement This study was in part supported by MIC SCOPE (27-J-J6238).

References

1. Chubb, E.C., Colgate, J.E., Peshkin, M.A.: Shiverpad: A glass haptic surface that produces shear force on a bare finger. IEEE Trans. Haptics **3**(3), 189–198 (2010)
2. Fagiani, R., Massi, F., Chatelet, E., Berthier, Y., Akay, A.: Tactile perception by friction induced vibrations. Tribol. Int. **44**(10), 1100–1110 (2011)
3. Imaizumi, A., Okamoto, S., Yamada, Y.: Friction sensation produced by laterally asymmetric vibrotactile stimulus. In: Auvray, M., Duriez, C. (eds.) Haptics: Neuroscience, Devices, Modeling, and Applications. Springer, Berlin (2014)
4. Konyo, M., Yamada, H., Okamoto, S., Tadokoro, S.: Alternative display of friction represented by tactile stimulation without tangential force. In: Ferre, M. (ed.) Haptics: perception, devices and scenarios. pp. 619–629. Springer, Berlin (2008)

5. Smith, A.M., Chapman, C.E., Deslandes, M., Langlais, J.S., Thibodeau, M.P.: Role of friction and tangential force variation in the subjective scaling of tactile roughness. Exp. Brain Res. **144** (2), 211–223 (2002)
6. Tappeiner, H.W., Klatzky, R.L., Unger, B., Hollis, R.: Good vibrations: asymmetric vibrations for directional haptic cues. In: Proceedings of IEEE World Haptics Conference, pp. 285–289 (2009)

Part III
Force Feedback Devices and Rendering

Pressure Threshold of the Hanger Reflex at the Wrist

Takuto Nakamura, Narihiro Nishimura, Taku Hachisu, Michi Sato and Hiroyuki Kajimoto

Abstract While hanger reflex is known as an involuntary movement of human head, the similar phenomenon has been found at the wrist and positions that efficiently generate the movement. However, the detailed condition about the strength of the pressure has not yet studied. In this paper, we measured the pressure thresholds of the hanger reflex at the wrist. The results showed around 6 N pressure presented both inward and outward rotation.

Keywords Hanger reflex · Pseudo-force · Force display · Wrist

T. Nakamura (✉) · N. Nishimura · T. Hachisu · M. Sato · H. Kajimoto
The University of Electro-Communications, 1-5-1 Chofugaoka, Chofu, Tokyo 182-8585, Japan
e-mail: n.takuto@kaji-lab.jp

N. Nishimura
e-mail: n-nishimura@kaji-lab.jp

T. Hachisu
e-mail: hachisu@kaji-lab.jp

M. Sato
e-mail: michi@kaji-lab.jp

H. Kajimoto
e-mail: kajimoto@kaji-lab.jp

T. Hachisu
JSPS, Tokya, Japan

H. Kajimoto
Japan Science and Technology Agency, Tokyo, Japan

H. Kajimoto et al. (eds.), *Haptic Interaction*, Lecture Notes in Electrical Engineering 277, DOI 10.1007/978-4-431-55690-9_22

1 Introduction

While conventional force feedback system takes large space and high cost, there were several attempts to present pseudo-force using perceptual illusion and thus to shrink space and lessen cost [1–4]. The hanger reflex is one of the perceptual pseudo-force illusions that a wire hanger worms by the head induces rotational force and involuntary rotation [5]. Sato et al. discovered that the pressure in the temporal region of the head is the trigger of this phenomenon and developed a device that controls the pressure and the generation of the hanger reflex [6]. In our previous report, we found that the similar phenomenon occurs in the waist and wrist [7]. In particular in the wrist, we discovered the positions that efficiently generate the hanger reflex, just like an ordinary hanger reflex at the head, and developed the device that controls the generation of the hanger reflex [8]. However, the necessary pressure value and its variation among people have not yet studied. In this paper, using previously developed wrist-type hanger reflex device, we investigated thresholds of the pressure that people perceive the twisting-force during the hanger reflex at the wrist.

2 Experiment

2.1 Setup and Participants

In the experiment, we used the previously developed twisting-force device [8] to generate hanger reflex at the wrist and measured the pressure thresholds. On the device, four linear actuators (Miniature Linear Motion series PQ12, Firgelli Technologies Inc.) are mounted and generate the hanger reflex by pushing the "sweet spots." Also, the force sensor (FSR 400, Interlink Electronics Inc.) is attached to each linear actuator and measured the force that the linear actuator applied to the user. Five laboratory members participated in this experiment (5 males, age range from 22 to 26 years old). Before the experiment, we confirmed that the hanger reflex occurs on their wrist, and let them remember the feeling of the twisting-force caused by the hanger reflex. During the measurement, the participants were equipped with the twisting-force device on their left wrist and kept standing with their left palm tuning inside. Also, they wore a sleep mask and listened to white noise with noise-canceling headphone to mask visual and auditory cues (Fig. 1).

2.2 Procedures

A method of limits was used to measure the thresholds of the pressure for the twisting-force caused by the hanger reflex. In the measurement, seven force levels

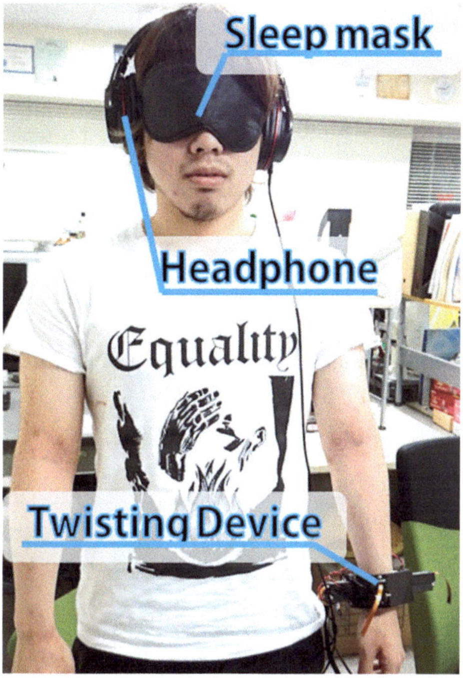

Fig. 1 A participant during the measurement

(4.43, 4.65, 5.63, 6.01, 6.81, 7.75, and 8.82 N) were prepared and applied in ascending or descending order by the twisting-force device until they started to feel the rotation or the felt the rotational force disappeared. For each participant, the measurement of outward rotation was conducted first, and inward rotation was conducted consecutively.

For each trial, we asked the participant to answer if he perceived the twisting-force or not (forced choice). We let them move their wrists to confirm their feeling. Hence, we obtained upper and lower thresholds for each condition, and the threshold was obtained by averaging the two.

2.3 Results and Discussion

Figures 2 and 3 show the average of upper and lower thresholds of each participant, and the average of them. In the outward rotation, the thresholds ranged 5.73 and 7.28 N, and the average and its standard deviation were 6.47 and 0.63 N, respectively. In the inward rotation, the thresholds ranged 5.03 and 7.28 N, and the average and its standard deviation were 5.86 and 0.92 N, respectively. The result indicates that the hanger reflex at the wrist can be generated by the commercial actuators. Because around 6 N can be generated by small actuators, the twisting-force device can be implemented in small size also.

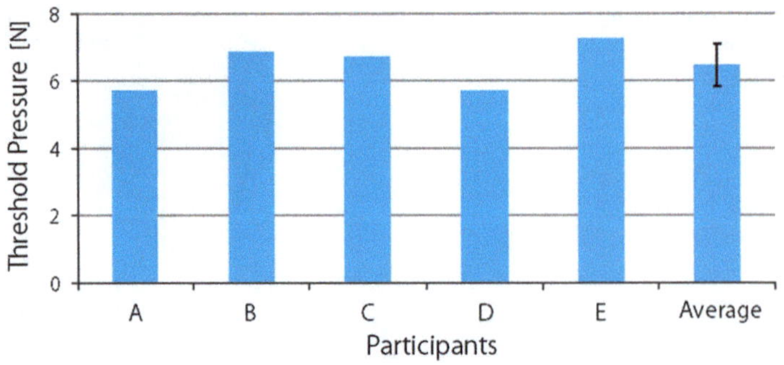

Fig. 2 Results in the measurement of the external rotation

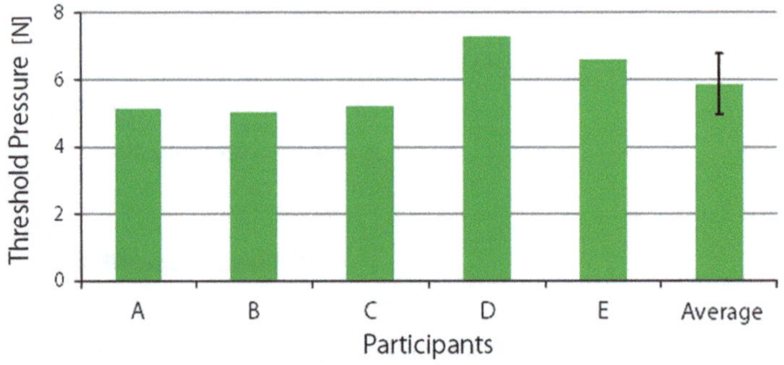

Fig. 3 Results in the measurement of the inner rotation

Comparing the averaged thresholds, the inward rotation had around 0.6 N lower thresholds than that of the outward rotation, but t test did not show significant difference between the two ($p = 0.12$). However, most participants commented that the inward rotation was easier to detect, which might be due to the initial posture.

3 Conclusion

In this paper, we measured the pressure thresholds of the twisting-force caused by the hanger reflex at the wrist. The results showed the hanger reflex needs around 6 N to generate, which can be presented by the commercial actuators. Comment from participants showed the inward rotation might have lower threshold than the outward rotation, but it might be due to the initial posture of the experiment.

Our future work will include measurement with different posture, more accurate thresholds measurement using a method of constant, and the measurement of the torque that the user perceive during this illusion. We will also work on revealing the mechanism of the illusion by closely observing skin deformation, bone deformation, and applied force.

References

1. Amemiya, T., Ando, H., Maeda, T.: Phantom-DRAWN: direction guidance using rapid and asymmetric acceleration weighted by nonlinearity of perception. In: 2005 International Conference. Augmented Tele-existence, pp. 201–208. ACM Press, New York (2005)
2. Rekimoto, J.: Traxion: a tactile interaction device with virtual force sensation. In: Proceedings of the 26th annual ACM symposium on user interface software and technology. ACM Press, New York (2013)
3. Minamizawa, K., Prattichizzo, D., Tachi, S.: Simplified design of haptic display by extending one-point kinesthetic feedback to multipoint tactile feedback. In: 2010 IEEE Haptics Symposium, pp. 257–260. IEEE Press (2010)
4. Solazzi, M., Provancher, W.R., Frisoli, A., Bergamasco, M.: Design of a SMA actuated 2-DoF tactile device for displaying tangential skin displacement. In: 2011 IEEE World Haptics Conference (WHC), pp. 31–36. IEEE Press (2011)
5. Matsue, R., Sato, M., Hashimoto, Y., Kajimoto, H.: Hanger reflex: a reflex motion of a head by temporal pressure for wearable interface. In: SICE Annual Conference, pp. 1463–1467. (2008)
6. Sato, M., Matsue, R., Hashimoto, Y., Kajimoto, H.: Development of a head rotation interface by using hanger reflex. In: The 18th IEEE International Symposium on Robot and Human Interactive Communication (RO-MAN), pp. 534–538. IEEE Press (2009)
7. Nakamura, T., Nishimura, N., Sato, M., Kajimoto, H.: Application of hanger reflex to wrist and waist. In: 2014 IEEE Virtual Reality (VR), pp. 181–182. IEEE Press (2014)
8. Nakamura, T., Nishimura, N., Sato, M., Kajimoto, H.: Development of a wrist-twisting haptic display using the hanger reflex. In: 11th Advances in Computer Entertainment Technology Conference. ACM Press (2014) (accepted)

SPIDAR-S: Haptic Device Attached to the Smartphone

Motonori Toshima, Katsuhito Akahane and Makoto Sato

Abstract In this study, we propose a haptic device SPIDAR-S which is able to be attached to the smartphone. Haptic devices that have been developed before are not familiar to many people because they are expensive, large, and complex. Therefore, we aimed to develop SPIDAR-S which is a very small and simple device. And we hope that many people will be able to enjoy haptic technology easily in any place by using it.

Keywords SPIDAR · Haptic device · Smartphone

1 Background

Recently, many haptic devices, such as PHANTom [1] and Falcon [2], have been developed. But these devices are not familiar to a lot of people. Possible reasons are that devices are expensive, large, and complex. So, we try to create a small and simple device, and many people will be able to enjoy the haptic technology. We paid attention to the smartphones. They have become widespread. Its features are high possession rate, high performance and multiple function, etc.

2 Spidar-S

2.1 Device Composition

SPIDAR-S is a very small and simple haptic device. SPIDAR-S provides force feedback using one thread and one motor. The direction of the force is in one

M. Toshima (✉) · K. Akahane · M. Sato
Precision and Intelligence Laboratory, Tokyo Institute of Technology, Tokyo, Japan
e-mail: toshima.m.aa@m.titech.ac.jp

H. Kajimoto et al. (eds.), *Haptic Interaction*, Lecture Notes in Electrical Engineering 277, DOI 10.1007/978-4-431-55690-9_23

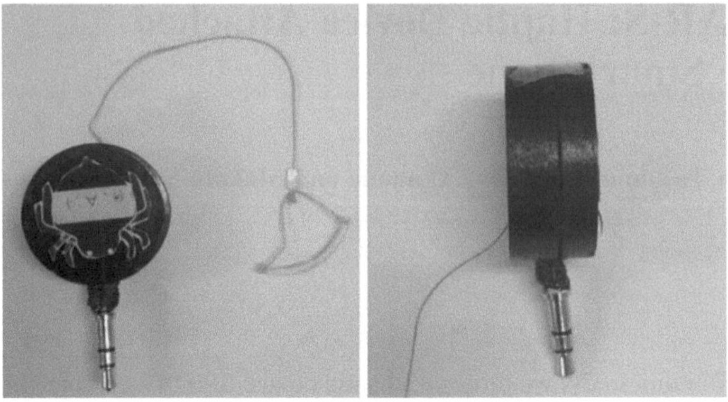

Fig. 1 SPIDAR-S overview

direction. And we adopted the earphone jack as an interface with the smartphone. Earphone jack is mounted on almost smartphones. So, SPIDAR-S is a versatile device.

2.2 *Motor Control*

The motor is controlled by the audio output from the smartphone. We amplify and smooth the audio signal from the smartphone because it is very small and AC signal. And we control the motor by using MOSFET as a switch. If the smartphone outputs the audio signal, motor is driven. On the other hand, if the smartphone does not output the audio signal, motor is stopped (Fig. 1).

2.3 *How to Use*

We explain how to use the device. User has a smartphone with one hand and moves the other hand in the back of the smartphone. SPIDAR-S detects user's finger position by using camera which is mounted on the smartphone. We put a marker in the thread, and it detects the marker position by the image processing. It provides force feedback by the image information.

3 Future Works

This time, we proposed a small haptic device SPIDAR-S which is attached to the smartphone. We need to improve its detection accuracy and review the force provided by the motor.

Acknowledgement We would like to thank Tetsuya Harada and Takehiko Yamaguchi for their helpful comments.

References

1. SensAble: http://www.sensable.com
2. Novint: http://www.home.novint.com

Pseudo-Haptic Interface Using Multipoint Suction Pressures and Vibrotactile Stimuli

Daiki Maemori, Lope Ben Porquis, Masashi Konyo and Satoshi Tadokoro

Abstract Dexterity for fine manipulation requires information from multiple skin contacts to detect the external forces applied on a tool. In this study, we developed a haptic interface to represent the external forces or stiffness of objects so that the pressure distributions at the contact pads can be controlled by using suction stimuli. The original interface had the drawback of being unable to represent high-frequency force sensations such as friction and collision because of air stimuli have large-scale characteristics. Thus, we integrated vibrotactile stimuli into the interface to represent high-frequency force sensations.

Keywords Haptic interfaces · Force illusion · Suction pressure · Vibrotactile stimuli · Tool manipulation

1 Introduction

People can perform tasks with dexterity when using various tools as if they are moving their own fingers. Dexterity can be achieved even if force information is indirectly sensed by the skin. Therefore, multiple skin contact cues are extremely important for tool manipulation.

D. Maemori (✉) · L.B. Porquis · M. Konyo · S. Tadokoro
Graduate School of Information Science, Tohoku University,
6-6-01 Aramaki Aza Aoba, Aoba-ku, Sendai, Miyagi, Japan
e-mail: maemori@rm.is.tohoku.ac.jp
URL: http://www.rm.is.tohoku.ac.jp

L.B. Porquis
e-mail: lopeben@rm.is.tohoku.ac.jp

M. Konyo
e-mail: konyo@rm.is.tohoku.ac.jp

S. Tadokoro
e-mail: tadokoro@rm.is.tohoku.ac.jp

H. Kajimoto et al. (eds.), *Haptic Interaction*, Lecture Notes in Electrical Engineering 277, DOI 10.1007/978-4-431-55690-9_24

2 Pen-Type Pseudo-Haptic Interface

We previously tried to represent external forces and torques by controlling the skin pressure distributions using suction stimuli at multiple contact areas on the skin [1]. There are several kinds of tactile stimuli, but we preferred to apply air suction to induce force sensations. Using an appropriate hole size to vacuum the skin can induce almost the same mechanoreceptor activity as pressing the skin [2]. Suction stimuli have the advantage of creating new interfaces that contain no moving parts. They are compact and can stabilize the contact between the skin and tool. For the interface prototype, we used two contact pads for the thumb and index to simply the control and distribution of the pressure stimuli. We divided each contact pad into four sections, and the pressure on each section can be controlled individually. Thus, the interface has a total of eight pressure chambers. Using eight pressure controllable chambers is enough to evoke force and torque sensations with up to six degrees of freedom (DOF).

We developed a suction pressure interface called the **T**actile **A**ugmented **K**inesthetic illusi**O**n Pen (TAKO-Pen) by combining the method for representing a force vector with a six-DOF force sensor [1] in a pen-type interface. Figure 1 shows an image of the TAKO-Pen. We also demonstrated that suction pressure can represent virtual stiffness by enhancing force perception. The perceived force is enhanced by the suction pressure augmenting the physical force. This demonstration is an example of augmented reality technology [1].

A significant drawback of the TAKO-Pen is its slow response, even if it is fairly reliable at representing the force vector or stiffness of an object. The slow response is caused by the characteristics of air stimuli. Because air is a compressive fluid, there is a time lag to reach the desired amount of pressure level. In addition, suction pressure is only effective at stimulating the slowly adapting type I (SA-I) mechanoreceptors, which are otherwise known as Merkel disks. The old display method of the TAKO-Pen was optimized to activate SA-I receptors other than mechanoreceptors (Meissner or Pacinian) that respond to high-frequency

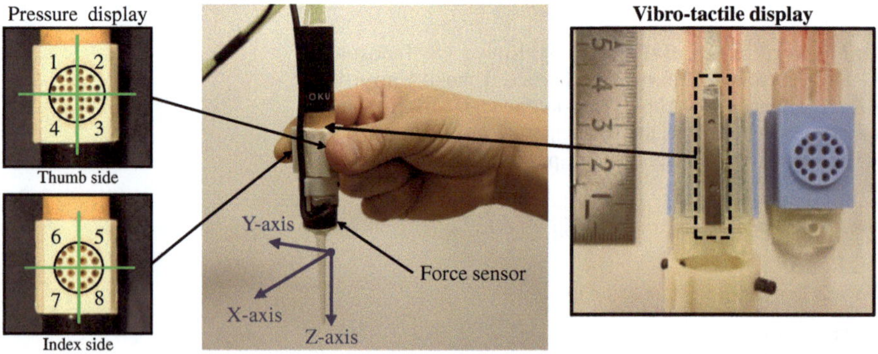

Fig. 1 Appearance of haptic interface called TAKO-Pen and each type of tactile display formation

information. Thus, the TAKO-Pen could not represent high-frequency information such as collisions when the tool collided with something or friction. In order to compensate for the slow response speed, we added a vibration actuator to the TAKO-Pen. We used a friction presentation method for 1-DOF vibrotactile stimuli [3]. This method produces mechanoreceptor activity (Meissner or Pacinian) at the time of the stick-slip transition. High-frequency vibrotactile stimuli greater than 200 Hz are presented the moment after the transition from sticking to slipping based on an approximation model of the stick-slip transition.

Acknowledgement This work was partly supported by JSPS KAKENHI Grant Number 26630083.

References

1. Porquis, L.B., Maemori, D., Nagaya, N., Konyo, M., Tadokoro, S.: Presenting virtual stiffness by modulating the perceived force profile with suction pressure. In: 2014 IEEE Haptics Symposium, pp. 289–294 (2014)
2. Makino, Y., Asamura, N., Shinoda, H.: A whole palm tactile display using suction pressure. In: Proceedings of 2004 IEEE International Conference on Robotics and Automation, pp. 1524–1529 (2004)
3. Konyo, M., Yamada, H., Okamoto, S., Tadokoro, S.: Alternative display of friction represented by tactile stimulation without tangential force. In: Proceedings of EuroHaptics 2008, pp. 619–629 (2008)

Wearable Pseudo-Haptic Interaction by Using Electrical Muscle Stimulation

Takaaki Ishikawa, Toshio Tsuji and Yuichi Kurita

Abstract Although electrical muscle stimulation (EMS) has been utilized in various haptic interfaces, haptic sense induced by EMS does not satisfactorily provide the full spectrum of haptic events. To enhance pseudo-haptic feedback provided via EMS, we incorporated visual feedback and examined the effect of such feedback on the pseudo-haptic feedback induced by EMS. Results showed that the perceived forces in the electrical stimulation with visual feedback were larger than the force exerted in the sole electrical stimulation. That is, it is suggested that the pseudo-haptic feedback can be enhanced and controlled by the addition of the visual feedback.

Keywords Haptic display · Electrical muscle stimulation · Visual feedback

1 Introduction

The field of haptics is growing: Haptic displays are now viewed as a highly promising approach to interfaces in human–computer interaction (HCI) tools. Haptic feedback systems are generally classified, according to their mechanical

---

T. Ishikawa (✉)
Graduate School of Engineering, Hiroshima University, 1-4-1 Kagamiyama, 739-8527 Higashihiroshima City, Hiroshima, Japan
e-mail: takaaki@bsys.hiroshima-u.ac.jp

T. Tsuji · Y. Kurita
Institute of Engineering, Hiroshima University, 1-4-1 Kagamiyama, 739-8527 Higashihiroshima City, Hiroshima, Japan
e-mail: tsuji@bsys.hiroshima-u.ac.jp

Y. Kurita
e-mail: kurita@bsys.hiroshima-u.ac.jp

Y. Kurita
JST PRESTO, 7 Gobancho, 102-0076 Chiyoda-ku, Tokyo, Japan

H. Kajimoto et al. (eds.), *Haptic Interaction*, Lecture Notes in Electrical Engineering 277, DOI 10.1007/978-4-431-55690-9_25

grounding configuration, into grounded systems, which include Phantom [1] and SPIDAR [2], and wearable systems, such as CyberGrasp [3] and Weight/Friction Illusion Display [4]. However, a problem of these device is that there is a limit to the area or position where is possible to display force. So, it is well known in the field that applying electrical pulses via electrode pads with controlled amplitude, timing, and frequency induces muscle contractions. If you use electrical stimulation, haptic display in various area or position is possible by changing the position of the electrodes. This electrical muscle stimulation (EMS) is utilized in haptic interfaces for mixed reality applications [5], interactive public displays [6], and induced motion illusions [7]. Kruijff et al. [8] successfully demonstrated pseudo-haptic feedback by using neuromuscular electrical stimulation (NMES) methods. They investigated the physiological processes of pseudo-haptic feedback and reported that in using NMES, it is difficult to provide the full spectrum of haptic events in arm movements, because stimulating an arm cannot generate sufficient muscle activity as compared to self-induced muscular activities performed voluntarily.

In addition to EMS, visual feedback also induces haptic sensations. Ban et al. [9] developed a method that induced illusionary effects in weight perception by changing visual properties of an object. Taking these studies into consideration, we utilized visual feedback in conjunction with EMS to enhance pseudo-haptic feedback. In this study, we examined the effects of such visual feedback on pseudo-haptic feedback induced by EMS.

2 Methods

2.1 Pseudo-Haptic Display via Electrical Stimulation

Figure 1 shows the overview of our system, which consisted of two parts: a visual display and EMS feedback. Visual information was displayed on a head-mounted display (HMD) by using an augmented reality method. We used the Unity engine

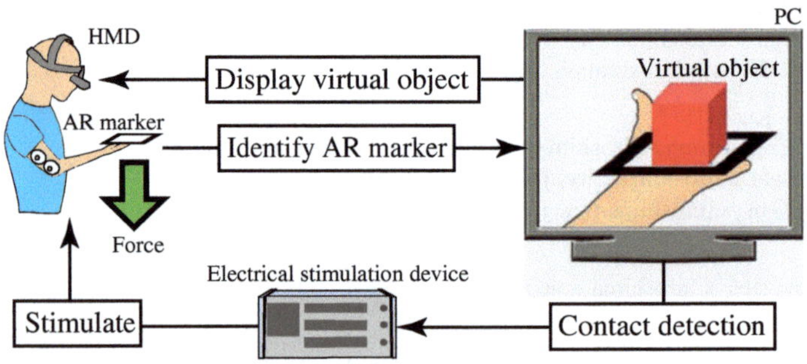

Fig. 1 Overview of our pseudo-haptic display system

[10] and the ARToolKit [11] to build a virtual object and create virtual interactions between the virtual object and a physical marker placed on a user's hand. In the user's view, a gray cube was displayed approximately 10 cm above the marker. The cube dropped when an experimenter gave a trigger signal.

In the experiment, electrical stimulation was given when the virtual cube reached the virtual plane where the physical marker was located. EMS feedback was produced by an EMS device (H2L Inc.; PossessedHand [12]) that was able to generate electrical pulses of up to 42 V. We were also able to adjust electrical pulse parameters via serial communication, such as the electrical path, electrical pulse height (voltage), electrical width, and electrical frequency. The applied stimulation pulse width was set to 200 μs and the pulse frequency was set to 40 Hz. PossessedHand stimulated the triceps brachii muscle of the user to generate a flexion moment at the elbow joint. This moment is as same as the moment acting on the arm when you hold the real object, and it creates an illusion as weight perception.

2.2 Experiment

Using this system, we verified how weight perception is changed by visual feedback. Figure 2 shows an overview of our experiment. Ten healthy male subjects with ages ranging from 22 to 24 participated in this study, each of them providing informed consent before participation. Subjects were instructed to flex their dominant arm at a 90° angle. Two surface electrodes were attached on the skin of the dominant arm where muscle contraction on the triceps brachii muscle is clearly induced. The non-dominant arm was placed on a desk, and a force transducer was put on the hand of the non-dominant arm (Fig. 3). Next, the experimenter gave

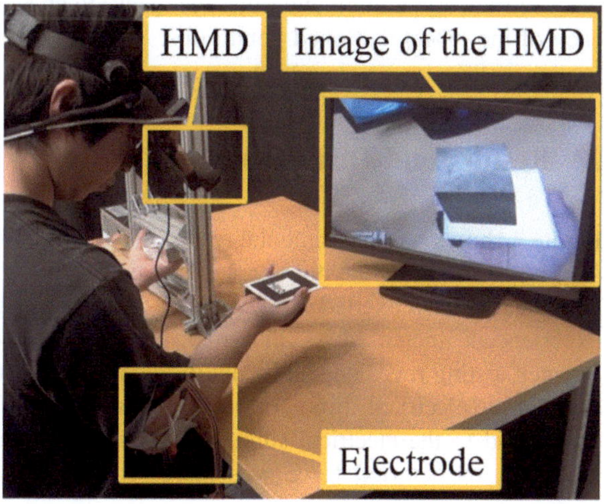

Fig. 2 Overview of the experiment

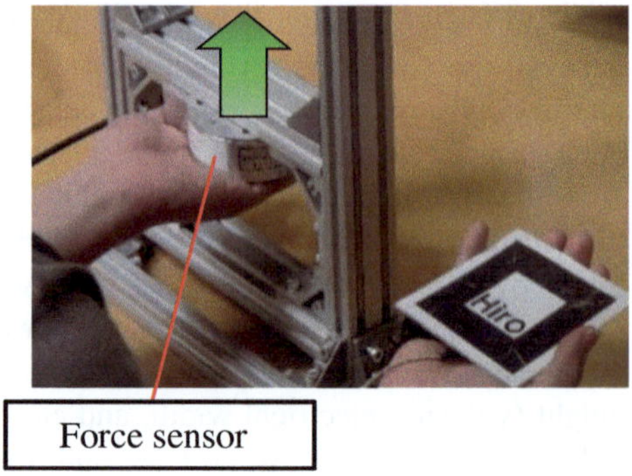

Fig. 3 Measurement of perceived force by EMS

electrical stimulation at the subject's dominant arm. While giving the electrical stimulation, the experimenter instructed the subjects to maintain their initial attitude and to reproduce the force they perceived by pushing the force transducer at the non-dominant arm. This method consults a Lee's method [13], which is evaluating the sensorimotor function. Thus, it is possible to regard a subjective value that the force they perceived by electrical stimulation as an objective value.

We performed the test about two conditions to verify the influence of visual feedback: (1) only electrical stimulation, (2) electrical stimulation with visual feedback. In the only electrical stimulation test, where visual feedback was not provided, the experimenter instructed the subjects to close their eyes and then stimulated the subject's muscles with voltage levels of 40, 41, and 42 V. In the electrical stimulation with visual feedback test, the experimenter instructed the subjects to equip a HMD and look at the image that was displayed in the HMD. The electrical stimulation was synchronized with the timing of the virtual object shown in the display. One side of the virtual object is approximately 10 cm, and pasted texture of metallic was displayed. For all of the above tests, the subjects were asked to exert a force upward to match the force they perceived from the combination of the electrical stimulation and other feedback. Each subject underwent three trials for each of the aforementioned conditions.

3 Results and Discussion

Figure 4 shows our experimental results. The vertical axis indicates the force that the subjects exerted, while the horizontal axis shows the intensity of the electrical stimulation. The exerted force, which is associated with the force the subjects experienced, increased as the intensity of the electrical stimulation increased. Dunnett testing revealed significant differences against the sole electrical stimulation case for the electrical stimulation with voltage level of 41 V ($p < 0.05$). When visual was added to the electrical stimulation, the subjects exerted more force. By visual feedback, the exerted force was increased by 30 %. Visual information is well

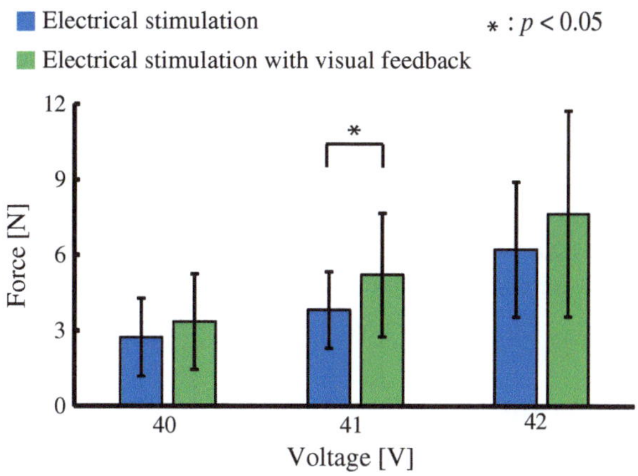

Fig. 4 Perceived force via EMS

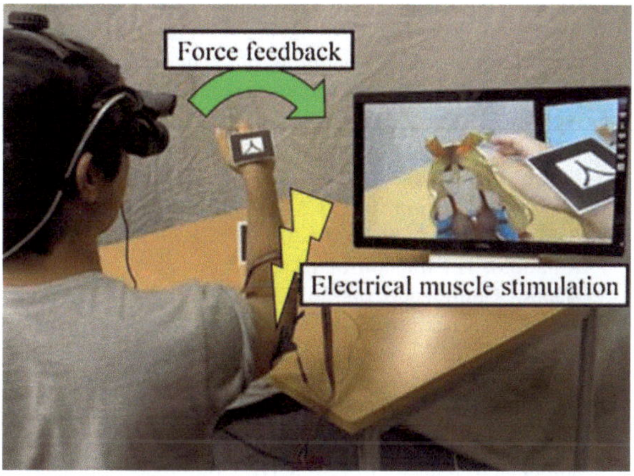

Fig. 5 Haptic interaction system

known to influence tactile and haptic perceptions. In our experiments, electrical stimulation was given when the virtual object contacted the virtual plane on which the physical marker was located. Because the physical marker was placed on the subject's palm, the visual feedback evokes a pseudo-contact sensation at the palm and therefore enhances the pseudo-haptic sensation from the electrical stimulation.

Based on these results, we developed a haptic interaction system with a virtual character by using EMS. A user attached electrodes for EMS and wore a HMD (Fig. 5). When the user's hand contacts with the virtual character, haptic feedback via EMS is generated. The synchronized stimulation of visual and electrical feedback realizes more intuitive interaction with a virtual object.

4 Conclusions

In this study, we developed a pseudo-haptic display system by using electrical stimulation. We examined the effect of visual feedback on the perceived force induced by EMS. Our experimental results showed that the perceived forces in the visual feedback were larger than the force exerted in the sole electrical stimulation test. Our future work includes exploring the feedback control technique to generate a desired force sensation.

References

1. Massie, T.H., Kenneth, S.J.: The PHANToM haptic interface: a device for probing virtual objects. In: The ASME Winter Annual Meeting, Symposium on Haptic Interfaces for Virtual Environment and Teleoperator Systems, pp. 295–300 (1994)
2. Kim, S., Hasegawa, S., Koike, Y., Sato, M.: Tension based 7-DOF force feedback device: SPIDAR-G. In: IEEE Virtual Reality, pp. 283–284 (2002)
3. CyberGrasp, Immersion Corp., http://www.immersion.com
4. Kurita, Y., Yonezawa, Y., Ikeda, A., Ogasawara, T.: Weight and friction display device by controlling the slip condition of a fingertip. In: IEEE/RSJ International Conference on Intelligent Robots and Systems, pp. 2127–2132 (2011)
5. Farbiz, F., Yu, Z.H., Manders, C., Ahmad, W.: An electrical muscle stimulation haptic feedback for mixed reality tennis game. In: ACM SIGGRAPH 2007, p. 140 (2007)
6. Pfeier, M., Schneegass, S., Alt, F.: Supporting Interaction in public space with electrical muscle stimulation. In: ACM Conference on Pervasive and Ubiquitous Computing Adjunct Publication 2013, pp. 5–8 (2013)
7. Kajimoto, H.: Illusion of motion induced by tendon electrical stimulation. In: IEEE World Haptics Conference, pp. 555–558 (2013)
8. Kruijff, E., Schmalstieg, D., Beckhaus, S.: Using neuromuscular electrical stimulation for pseudo-haptic feedback. In: ACM Symposium on Virtual Reality Software and Technology, pp. 316–319 (2006)
9. Ban, Y., Narumi, T., Fujii, T., Sakurai, S., Imura, J., Tanikawa, T., Hirose, M.: Augmented endurance: controlling fatigue while handling objects by affecting weight perception using augmented reality. In: SIGCHI Conference on Human Factors in Computing Systems, pp. 69–78 (2013)
10. Unity, Unity Technologies, http://www.unity3d.com
11. Kato, H., Billinghurst, M.: Marker tracking and HMD calibration for a video-based augmented reality conferencing system. In: 2nd International Workshop on Augmented Reality, Proceedings. pp. 85–94 (1999)
12. Tamaki, E., Miyaki, T., Rekimoto, J.: PossessedHand: techniques for controlling human hands using electrical muscles stimuli. In: SIGCHI Conference on Human Factors in Computing Systems, pp. 543–552 (2011)
13. Walsh, L.D., Taylor, J.L., Gandevia, S.C.: Overestimation of force during matching of externally generated forces. J. Physiol. **589**, 547–557 (2011)

Normal and Tangential Force Decomposition and Augmentation Based on Contact Centroid

Sunghoon Yim, Seokhee Jeon and Seungmoon Choi

Abstract This study presents a simple but effective approach to extract and selectively augment normal and tangential force components from a reaction force when a user interacts with a real object using a probe. The approach first approximates the behavior of a contact area with a single representative point, i.e., contact centroid. We use it to decompose the reaction force into a normal and a tangential force component. For augmentation, the two components are selectively amplified or diminished. For demonstration, we applied the approach in a breast tumor palpation scenario, where inhomogeneity due to hard nodules is amplified for better detectability of abnormality.

Keywords Augmentation · Palpation · Inhomogeneity · Contact centroid

1 Introduction

In haptic augmented reality (AR), real touch sensation is augmented by virtual stimuli, allowing us to alter the haptic properties of real objects [1]. Our current research examines the feasibility of this topic, with specific focus on the augmentation of real object's stiffness and friction. In our earlier work, we introduced a haptic AR system that can alter the stiffness and friction of a physical object [1]. In that work, preprocessing steps including geometry and dynamics modeling of a real environment were employed in order to construct models for real-time estimation of required values for augmentation, e.g., surface normal direction. The amount of

S. Yim · S. Choi (✉)
Pohang University of Science and Technology (POSTECH), Pohang, Republic of Korea
e-mail: choism@postech.ac.kr

S. Yim
e-mail: algorab@postech.ac.kr

S. Jeon
Kyung Hee University, Yongin, Republic of Korea
e-mail: jeon@khu.ac.kr

H. Kajimoto et al. (eds.), *Haptic Interaction*, Lecture Notes
in Electrical Engineering 277, DOI 10.1007/978-4-431-55690-9_26

preprocessing steps was tried to be minimized in order to preserve the advantage of AR: efficient construction of immersive environment without extensive modeling.

This study further increases the usability by removing all the preprocessing and modeling. Underlying approach is to utilize the idea of the contact centroid. Bicchi et al. approximated the dynamics of an area contact by introducing "contact centroid": a single representative point lying on a probe tip surface [2], which can be efficiently estimated in real-time using a F/T sensor at the tool. For augmentation, we calculate the contact centroid and used it for estimating geometrical information of contacting surface, e.g., surface normal and tangential directions. Further, we use the directions for decomposing the force due to stiffness and the force due to friction. The decomposed forces are then used to selectively augment stiffness or friction.

2 Force Decomposition and Augmentation

Our framework uses a rigid spherical probe tip instrumented with a 6 axis F/T sensor (NANO17, ATI Technologies; see Fig. 1).

Ideally, in order to extract the normal force component from an area contact, normal forces should be sampled and integrated over the whole contact area, which needs a sophisticated sensor array [3]. Alternatively, we assume that the characteristics of the traction distribution over a contact area can be approximated using a single representative contact point, i.e., contact centroid [2], a reaction force, and torque acting on the point. A contact centroid is defined as a point on the probe tip surface where the resultant moment acting on it is parallel to the surface normal at that point, which can be easily estimated using an algorithm in [2].

Since the contact centroid is a point on the spherical probe tip, the contact normal direction can be easily found: direction from the contact centroid to the center of the probe tip. Using this direction, the reaction force vector can be

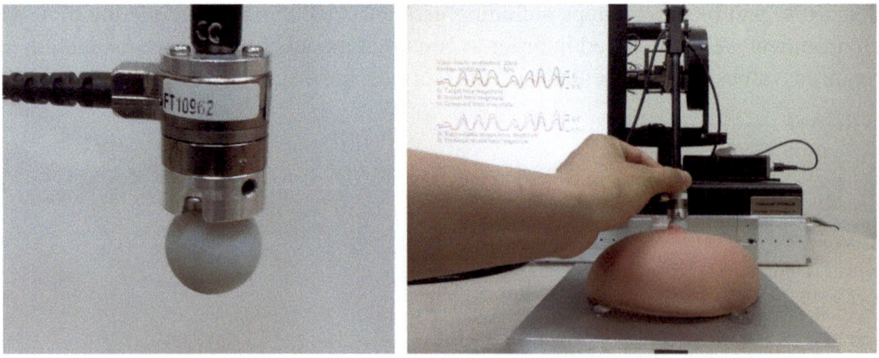

Fig. 1 A probe with a F/T sensor (*left*) and our haptic AR system (*right*)

decomposed into two perpendicular force vectors, i.e., normal $\mathbf{f}_{normal}$ and tangential forces $\mathbf{f}_{tangential}$. Finally, the force modulation is performed by exerting additional force:

$$\mathbf{f}_{command} = (k-1)\mathbf{f}_{normal} + (j-1)\mathbf{f}_{tangential}, \quad (1)$$

where k and j are amplification gains for normal and tangential modulation, respectively.

We tested the framework using a breast mock-up with hard nodules inside (see Fig. 1). According to our pilot user test, subjects are likely to detect more nodules when the normal force is amplified. During our demonstration in AsiaHaptics, we plan to conduct a more extensive user study to reveal the effect of augmentation on nodule detection.

Acknowledgements This work was supported in part by a Grant 2010-0019523, 2013R1A2A2A01016907, 2011-0027994, and 2011-0030075 all from NRF, and by an ITRC program (NIPA-2010-C1090-1011-0008) from NIPA, all funded by the Korean government.

References

1. Jeon, S., Choi, S.: Real stiffness augmentation for haptic augmented reality. Presence Teleoperators Virtual Environ. **20**(4), 337–370 (2011)
2. Bicchi, A., Salisbury, J.K., Brock, D.L.: Contact sensing from force measurements. Int. J. Robot. Res. **12**(3), 249–262 (1993)
3. Kim, H., Choi, S., Chung, W.K.: Contact force decomposition using tactile information for haptic augmented reality. In: Proceedings of IEEE/RSJ International Conference on Robots and Systems (IROS), pp. 1242–1247 (2014)

Object Manipulation by Deformable Hand

Koichi Hirota, Yusuke Ujitoko, Kazuya Kiriyama and Kazuyoshi Tagawa

Abstract This study investigated the implementation of a deformable elastic hand model. We used the finite element method (FEM) and the penalty method for physics-based computations of forces and deformations on the contact area. To accelerate the computation, we developed an efficient method that utilizes the link structure of the hand. Using this model, users can perform object manipulation and get visual feedback from the deformations in the skin.

Keywords Hand model · Deformation · Manipulation

1 Background

Manipulation of virtual objects using a human hand model has played an important role in recent virtual reality systems. Our goal was to manipulate virtual objects dexterously using a deformable hand model. Various methods of modeling human

---

K. Hirota (✉)
Graduate School of Information Systems, The University of Electro-Communications, 1-5-1 Chofugaoka, Chofu 182-8585, Tokyo, Japan
e-mail: hirota@vogue.is.uec.ac.jp

Y. Ujitoko
Graduate School of Interdisciplinary Information Studies, The University of Tokyo, 7-3-1 Hongo, Bunkyo-ku, Tokyo, Japan
e-mail: yusuke.ujitoko@gmail.com

K. Kiriyama
Department of Mechanical Engineering, The University of Tokyo, 7-3-1 Hongo, Bunkyo-ku, Tokyo, Japan
e-mail: kazuya.kiriyama11@gmail.com

K. Tagawa
Ritsumeikan Global Innovation Research Organization, Ritsumeikan University, 1-1-1 Nojihigashi, Kusatsushi, Shiga, Japan
e-mail: tagawa@cv.vi.ritsumei.ac.jp

H. Kajimoto et al. (eds.), *Haptic Interaction*, Lecture Notes in Electrical Engineering 277, DOI 10.1007/978-4-431-55690-9_27

hand have been proposed [1, 2]. We consider that the deformable hand model has several advantages over solid hand models. First, the deformation of flesh enhances the senses of visual reality. Second, the accuracy of computation during interaction can be improved. When we grasp or manipulate an object with our hand, the hand deforms to adapt to the shape of the object and forms a contact area between the hand and the object. The difference in the contact area affects the frictional properties of the interaction, and it leads to the difference in behavior of the object. Third, the stress state of each element inside the hand can be determined, which will enable us to design a haptic interface that provides additional information such as pain sensation.

2 Hand Model

Our hand model consists of skeletons and flesh. We modeled the skeleton of the hand using 23 rigid bodies assembled with fixed nodes and 15 joints. The flesh consisted of a set of free nodes, which were separated into groups by the nearest rigid body. The volume of the flesh was modeled as a tetrahedral mesh; the model consisted of 32,471 nodes (including 24,885 free nodes) and 153,018 tetrahedral elements. Finite element method (FEM) and the penalty method were used for the physics-based computation. The equations were solved by the iterative method to completely utilize the link structure of the hand. To reduce the processing time, GPUs were utilized for the computation.

3 Implementation and Evaluation

Our system was composed of a PC (Intel Xeon 3.16 GHz) with GPUs (nVIDIA GTX780Ti and Quadro K4000), a data glove (5DT Ultra 14), a magnetic tracker (Polhemus Fastrak), and a stereo scopic display (nVIDIA 3D Vision glasses + LCD) (Fig. 1). The position and orientation of the hand and head were measured by the tracker, and the shape of the hand measured by the data glove was mapped to the joint angles in the hand model. The computation of the FEM was primarily performed by GTX780Ti, and the computation of stress and visual rendering by Quadro K4000.

It was confirmed that various types of manipulation, such as pinching and grasping, can be performed using the proposed hand model (Fig. 2). The cycle time of computation in most part of manipulation was below 10 ms (Fig. 3).

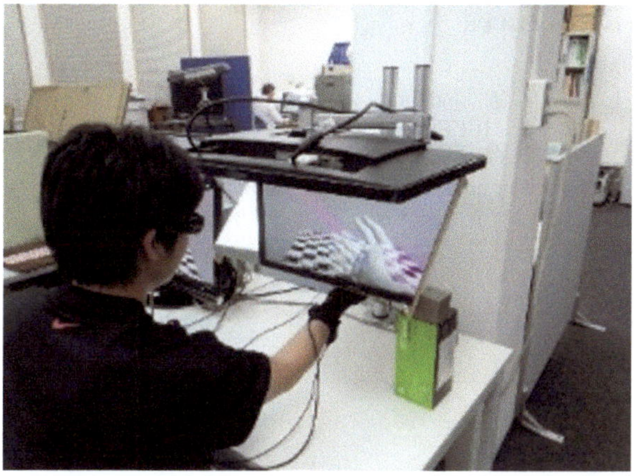

Fig. 1 Object manipulation environment

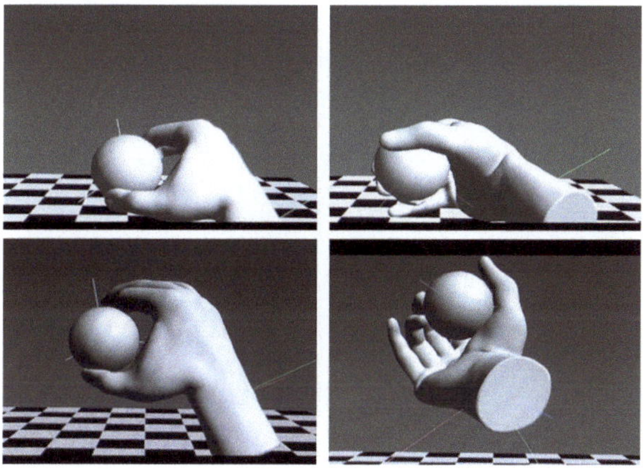

Fig. 2 Manipulating virtual objects

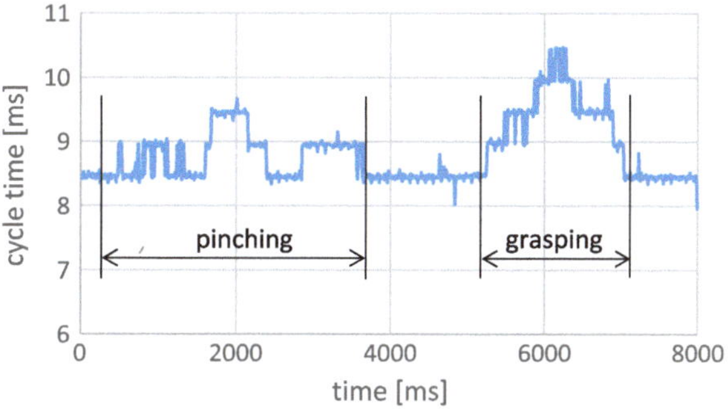

Fig. 3 Computation cycle time in object manipulation

References

1. Borst, C.W., Indugula, A.P.: Realistic virtual grasping. In: Virtual Reality, 2005. Proceeding. VR 2005. IEEE, pp. 91–98 (2005)
2. Perez, A.G., Cirio, G., Hernandez, F., Garre, C., Otaduy, M.A.: Strain limiting for soft finger contact simulation. In: World Haptics Conference (WHC), 2013, pp. 79–84 (2013)

A Proposal of Wire-Driven Bimanual Multi-finger Haptic Display SPIDAR-10

Hiroshi Koganeyama, Satoshi Miyake, Lanhai Liu, Naoki Maruyama, Katsuhito Akahane and Makoto Sato

Abstract In this research, we propose a system which builds a VR world where operation is possible like real world by using wire-driven bimanual multi-finger haptic display SPIDAR-10. By using SPIDAR-10, it is possible to present a haptic such as a sense to catch an object, gravity to depend on an object, and the power between the both hands through the object in a VR world, and users can manipulate virtual objects with an operation close to the operation of the real world.

Keywords SPIDAR-10 · Haptic device · Multi-finger · Hands

1 Introduction

1.1 Research Background

Real-time simulation of VR world has been made possible by improvements in computer performance. Accordingly, the demand of VR simulations that require complex operation such as surgery simulation, assembly simulation by using both hands is also increasing. From such a background, the wire-driven bimanual multi-finger haptic display SPIDAR-10 [1] that allows the fine technical operation by finger and twist of the hand has been developed in this laboratory.

1.2 Research Purposes

We provide an operating environment close to a real world. Therefore, we display a hand shape model based on ten fingertip positional information in the VR world.

H. Koganeyama (✉) · S. Miyake · L. Liu · N. Maruyama · K. Akahane · M. Sato
Precision and Intelligence Laboratory, Tokyo Institute of Technology, Yokohama, Japan
e-mail: koganeyama.h.aa@m.titech.ac.jp

H. Kajimoto et al. (eds.), *Haptic Interaction*, Lecture Notes
in Electrical Engineering 277, DOI 10.1007/978-4-431-55690-9_28

2 Device Configuration

SPIDAR-10 has two cylindrical rotary frames. Each frame is connected to the fixed frame by pulleys, and it is possible to rotate in the one axial direction by using a DC motor connected to the fixed frame. There are 5 finger-caps to put the fingers, and each finger-cap is connected to four wires (Fig. 1a, b).

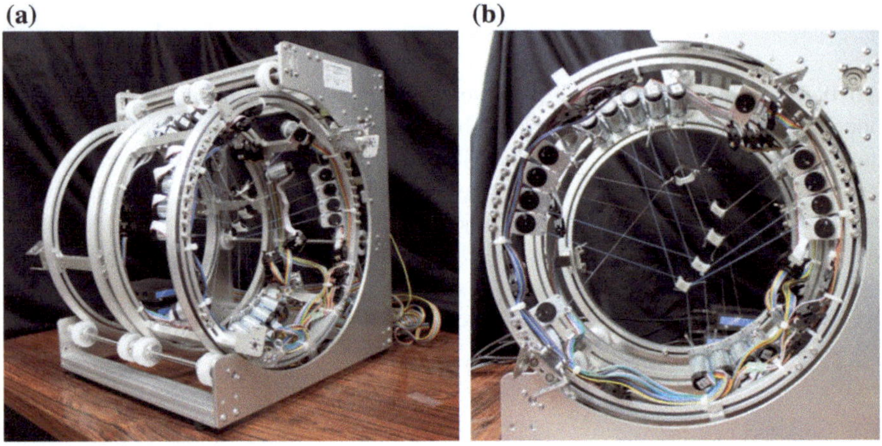

Fig. 1 SPIDAR-10 overview

Fig. 2 Hand shape model

3 Hand Shape Estimation

In this research, we make a 3-D hand shape model by the combination of sphere object and capsule object in a VR world (Fig. 2).

Reference

1. Shuto, M., Akahane, K., Wakita, W., Tanaka, H., Sato, M.: A proposal of a haptic interface for technical manipulation. In: PVRSJ The 17th Annual Conference, pp. 143–146 (2012)

Force Control of Stuffed Toy Robot for Intention Expression

Nutnaree Kleawsirikul, Yuanyuan Li and Shoichi Hasegawa

Abstract We propose a novel stuffed-toy robot with ability to express its intention through different responding force of the arms under force control. The robot is soft enough that people may want to embrace. The core of the research is the new force control mechanism for detection of the external force acting on the arms of the robot. The robot can sense the external force by a specialized force sensor and then change its stiffness with force control to respond to us. Different compliance on the arms can be realized by the change of force control gain.

Keywords Human–computer interaction · Robots · Force control

1 Introduction

Many stuffed-toy robots presented in the world today usually are constructed with hard mechanism as its movable parts. This is in contrast to Harlow's [1] study which suggests that softness and good hand feeling are important in developing affectionate interaction. Therefore, we propose a stuffed-toy robot which can express its intention, such as being peace or forced, through touch and force sensation on the arms of the robot while keeping its soft mechanism. We will present and demonstrate this robot to entertainment technology's community [2]. However, demonstration to and feedbacks from haptics community will also be valuable.

N. Kleawsirikul (✉) · Y. Li · S. Hasegawa
Hasegawa Shoichi Laboratory, Tokyo Institute of Technology, R2-624,
4259 Nagatsuta, Midori Ward, Yokohama, Kanagawa, Japan
e-mail: cnk@haselab.net
URL: http://haselab.net/

H. Kajimoto et al. (eds.), *Haptic Interaction*, Lecture Notes
in Electrical Engineering 277, DOI 10.1007/978-4-431-55690-9_29

2 Force Control Interaction for Intention Expression of Stuffed-Toy Robot

The stuffed-toy robot (Fig. 1a) developed based on previous work by Shiina and Hasegawa [3] can interact with users through force and touch sensation. Its structure is made up of soft textures such as clothes and cotton wools. Its movable parts, such as arms and head of the robot, are made by strings and cotton bags. The strings are controlled by motors to achieve the movement of the robot's arms and legs.

A special force sensor, which makes up of 2 photoreflectors in the slit on the duralmin body, can detect the external force acting on the arms in 2 directions based on the width of the slit. The force sensor is installed between the arm unit and a body side such that the movement of the arms will not interfere with the detection of external force.

As the external force is applied to the robot's arm, the analog-to-digital convertor (ADC) values are detected by the force sensor and converted into force value, and these values then are used to calculate the amount of changes in the string length for the next motion step. The amount of movement depends on the amount of force detected. The degree of compliance of the robot can be set by the amount of force control gain.

By this, the robot will be able to detect the external force applied to its arm by force sensor and provide force feedback through interactions with its arm based on force control method to express its intention (Fig. 1b).

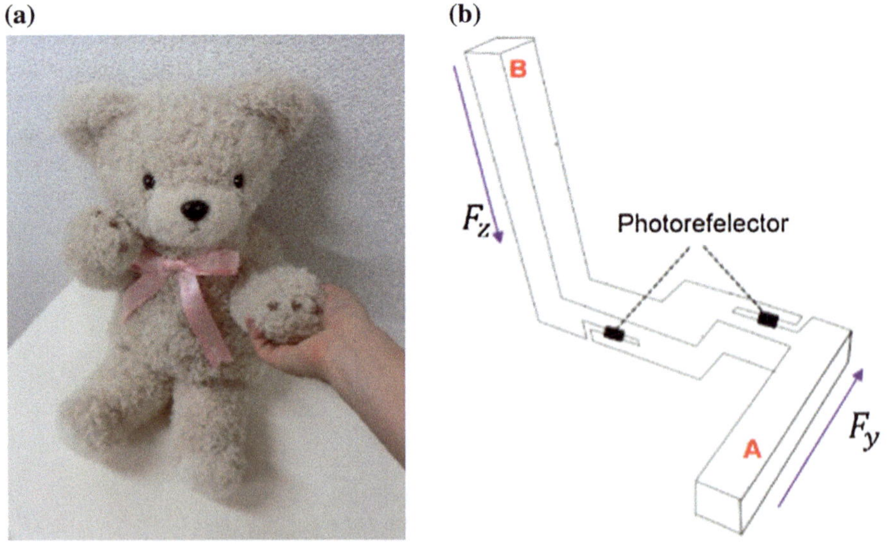

Fig. 1 The robot and the force sensor used in the result. **a** Our stuffed-toy robot. **b** The proposed force sensor

3 Conclusion

To evaluate the force control interaction, we have done an experiment. The result indicates the users' impression of the stuffed-toy robot can be changed by different force gain. Hence, it is possible to realize the robot's intention expression through force control.

References

1. Harlow, H.F.: The nature of love. Am. Psychol. **13**, 673–685 (1958)
2. Li, Y., Kleawsirikul, N., Takase, Y., Mitake, H., Hasegawa, S.: Intention expression of stuffed toy robot based on force control. In: Advances in Computer Entertainment Technology 2014, 11–14 Nov 2014
3. Shiina, M., Hasegawa, S.: Proposal of flexible string actuation mechanism for arm and leg of stuffed animal robot with good sense of touch. The Robotics and Mechatronics Conference **08** (4), 103 (2008)

Wearable 3DOF Substitutive Force Display Device Based on Frictional Vibrotactile Phantom Sensation

Ryota Nakagawa and Kinya Fujita

Abstract We have developed a substitutive three-degree-of-freedom force display device that is wearable on the fingertips. This device utilizes vibrotactile phantom sensation (VPS) and substitutes normal and tangential forces by using the VPS magnitude and position displacement. The device has five voice-coil actuators that produce oscillatory friction force to generate VPS. The adoption of friction force oscillation realized 8.5 mm thickness at the finger pad and 13.2 g weight.

Keywords Substitutive force display · 3DOF · Phantom sensation

1 Introduction

To achieve precise object manipulation in a virtual environment, it is essential to recognize both normal and tangential forces that are applied to the fingertips through the grasping object. Several types of fingertip-mounted pseudo-force display devices have been proposed to obtain a greater range of motion with a simple device structure. Although a wearable fingertip tightening device [1, 2] has been realized, its display information was limited within normal force. Therefore, we have proposed a substitutive force display device that utilizes the vibrotactile phantom sensation (VPS). This device substitutes the normal and tangential forces by the magnitude and position of the VPS [3]. However, a disadvantage of this device was its thickness, which disrupted the finger motion. In this study, we have developed a new device that utilizes friction force oscillation to generate VPS and thus achieve a less thick, lightweight, and easy-to-wear simple structure.

R. Nakagawa (✉) · K. Fujita
Graduate School, Tokyo University of Agriculture and Technology, 2-24-16 Nakacho, Koganei, Tokyo 184-8588, Japan
e-mail: 50014648128@st.tuat.ac.jp

K. Fujita
e-mail: kfujita@cc.tuat.ac.jp

H. Kajimoto et al. (eds.), *Haptic Interaction*, Lecture Notes in Electrical Engineering 277, DOI 10.1007/978-4-431-55690-9_30

2 Fingertip-Wearable 3DOF Substitutive Force Display Device

The principal concept of the proposed device is to substitute a force by controlling the VPS that is induced at the finger pad. Even though the VPS does not stimulate proprioceptive receptors and does not induce an actual sense of force, it does stimulate cutaneous receptors and has an ability to display three-degree-of-freedom information, namely the magnitude and the two-dimensional position. The normal force is substituted by the VPS magnitude, and the tangential force is substituted by the VPS position. Wearing the 3DOF substitutive force display device on the fingertips enables the display of both the parallel force and the moment applied to the hand.

To control the position of the VPS at the finger pad, four vibration actuators were utilized in the previous device. One problem with this design was the oscillation direction. Oscillation perpendicular to the finger pad increased the device thickness to 20 mm. To reduce the thickness, we adopted friction force oscillation. This design allowed the voice-coil actuators to be installed parallel to the finger pad, as shown in Fig. 1. The voice-coil actuator consists of a coil and a magnet embedded with a guiding shaft, as shown in Fig. 1b. The vibration is applied to the finger pad through a vibrating pin adhered to the coil surface. The problem caused by the use of frictional force oscillation is that the intensity of the vibration transmitted to the finger pad changes depending on the contact pressure between the pin and the finger pad. To obtain stable contact pressure, a pin height adjustment mechanism that pushes up the shaft of the voice-coil actuator by a setscrew was applied. Thereby, the thickness at the finger pad was reduced to 8.5 mm, as shown in Fig. 2.

Another problem was the resolution of the VPS position. To control the VPS position by using four vibrators located at the ends of a square, the ratio of the vibration magnitudes are controlled. This means that all four actuators are driven all

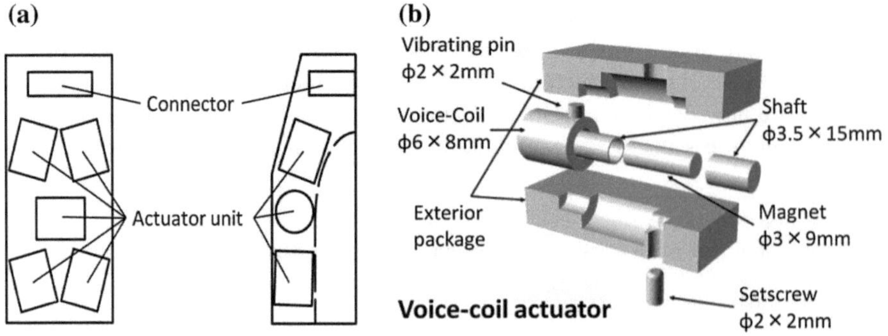

Fig. 1 **a** Internal structure of the developed 3DOF substitutive force display device and **b** internal structure of the voice-coil actuator unit installed in the device

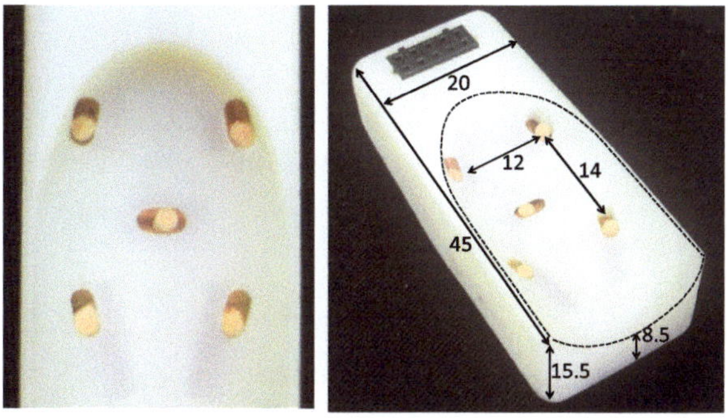

Fig. 2 Appearance of the developed 3DOF substitutive force display device

the time. The concern with driving all the vibrators is that it might induce VPS at a wider area of the finger pad and make the VPS position ambiguous. Therefore, we added another vibrator at the center of the square. This modification allows for not driving two vibrators at a time, because only three out of five vibrators are required to generate VPS at a specific position.

The weight of the developed device was 13.2 g. The simple structure attained easy d_{on} and d_{off} within several seconds. We have experimentally evaluated the direction recognition performance. The results demonstrated that the average recognition error of the VPS direction was reduced to 10°, whereas it was 25° with the previous device.

3 Conclusions

We have developed a substitutive 3DOF force display device that is wearable on the fingertips. By adopting friction force oscillation, 8.5 mm thickness at the finger pad and 13.2 g weight were achieved. The device wearing time was also reduced to several seconds. The device is expected to be applied to precise VR fingered-manipulation with less user load.

References

1. Inaba, G., Fujita, V.: A pseudo-force-feedback device by fingertip tightening for multi-finger object manipulation. In: Proceedings of EuroHaptics 2006, pp. 275–278 (2006)
2. Minamizawa, K., Tojo, K., Kajimoto, H., Kawakami, N., Tachi, S.: Haptic interface for middle phalanx using dual motors. In: Proceedings of EuroHaptics 2006, pp. 235–240 (2006)
3. Ooka, T., Fujita, K.: Virtual object manipulation system with substitutive display of tangential force and slip by control of vibrotactile phantom sensation. In: Proceedings of 2010 Haptics Symposium, pp. 215–218 (2010)

Proposal of 6 DoF Haptic Interface SPIDAR-I Optimized by Minimizing Margin of Peak Force

Yunong Ji, Hiroyuki Tajima, Katsuhito Akahane and Makoto Sato

Abstract In this paper, we describe the design of a wire-driven haptic device. One of the problems of wire-driven haptic devices is that peak force feedback are different by directions. If we solve this problem, we will be able to consider the design of wire-driven haptic device by minimal actuator output. Using min-max method, we sought the optimal geometric structure for omnidirectional isotropic peak force feedback and made a prototype according to it.

Keywords Haptic device · Wire driven · Min-max method · Isotropic peak force

1 Introduction

1.1 Background and Objective

Interface for 3D object manipulation, particularly which providing haptic interactions between user and object, is believed to be able to increase the sense of immersion to the VR environment, improve the operation efficiency, and therefore is extensively researched and developed [1, 2]. Our laboratory are conducting researches on parallel wires structured haptic interface SPIDAR [3] to realize force feedback, and various derivative devices have been developed to match specific purposes. One of them is a small-sized grip-type 6 DoF haptic interface specialized in desktop environment.

Studies for enhancing the performance of SPIDAR from various viewpoints have been made. About the optimality geometric structure, Igarashi developed a structure that maximized the calculation accuracy of position, posture, and force

Y. Ji (✉) · H. Tajima · K. Akahane · M. Sato
Precision and Intelligence Laboratory, Tokyo Institute of Technology, Yokohama
Kanagawa 226-8503, Japan
e-mail: ji.y.aa@m.titech.ac.jp

H. Kajimoto et al. (eds.), *Haptic Interaction*, Lecture Notes
in Electrical Engineering 277, DOI 10.1007/978-4-431-55690-9_31

distribution. And same as Igarashi, Koyama developed a structure that the values of peak force[1] are equal in the directions of three principal axes.

Due to the mechanism of realizing force feedback to users by the resultant of the tension of wires, the peak force of SPIDAR is different in different directions. Therefore, in case of realizing a force of a certain magnitude, although it is feasible in directions the peak force of which is large enough, the problem is that it is infeasible in directions the peak force of which is not. This means that it must be considered to ensure the peak force to be large enough in all directions when designing the haptic interface. Since it leads to the utilization of output of actuators without waste, it is helpful when designing the haptic interface to develop a structure of which the deviation of peak force is as small as possible.

2 Inner Strings Haptic Interface SPIDAR-I

2.1 Overview

SPIDAR is a parallel wire-driven haptic device capable of position, posture measurement, and force feedback. Features of SPIDAR are high rigidity, lightweight, smaller than other mechanisms, and high transparency since one of the components are wires.

As a haptic device designed for the desktop environment, SPIDAR-I has superb characteristics as follows.

- Being compact for wires that are disposed inside the grip
- Superb maneuverability and easy to hold the grip since wires and users' hands do not contact with each other
- Able to output moment efficiently compared with existing type SPIDAR-G

3 Design for Optimal Structure of SPIDAR-I

3.1 Device Specifications

Specifications of the device are set as follows.

- When the grip is in the home position, each wire is disposed in angular parameters as shown in Table 1, compared with prior types, where ϕ, θ, and ξ are structure parameters shown in Fig. 1.
- SPIDAR-I is a device that is intended to be used in a desktop environment, and a compact design is needed.

[1]Peak force: the largest force the end effector (hereinafter referred to as grip) of haptic device can exert on user.

Table 1 Comparison of structure parameters (unit: degree) and minimal peak force

	Igarashi type	Koyama type	Proposal type
ϕ	45.000	45.000	45.000
θ	35.264	42.941	42.013
ξ	0.000	0.000	0.000
$\|f_{\text{min-peak}}\|$	0.88135	0.88993	0.89013

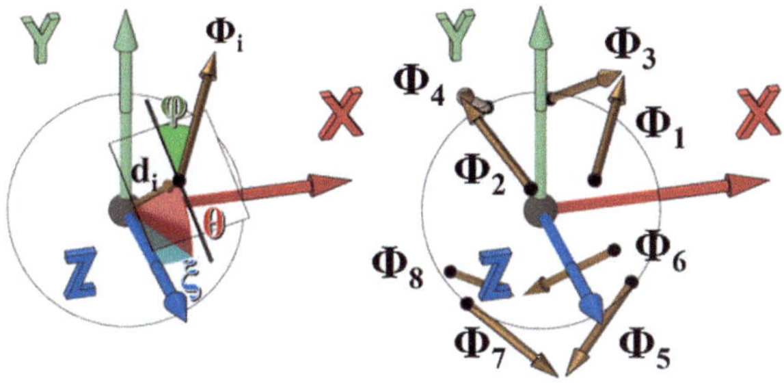

Fig. 1 Structure parameters and wire disposition

3.2 Design and Fabrication of Device

In this section, we designed a device that meets the specifications above. Radius of the grip R affects the range of movement of the grip. Therefore, it is desirable to set R as large as possible. Assume that the radius large enough to fit in the hand of an adult male was 44 mm. In order to make the device compact, we choose to place the 8 motors under the grip instead of placing 4 under and 4 above the grip, extend wires of 4 motors among them upward to the top of the device, and pull each wire through 2 guide rollers (Fig. 2).

SPIDAR-I generates tension by winding the wire using a bobbin which is mounted directly on the pivot of motor. Maximum continuous torque of the motor is 10.6 mN m. Since the inner diameter of bobbin is 10 mm, maximum continuous tension of the wire is 2.12 N under the premise of using within maximum continuous torque. To control this device, we use the controller that can realize haptic presentation with a high speed of 1 kHz. The device actually fabricated is shown in Fig. 3.

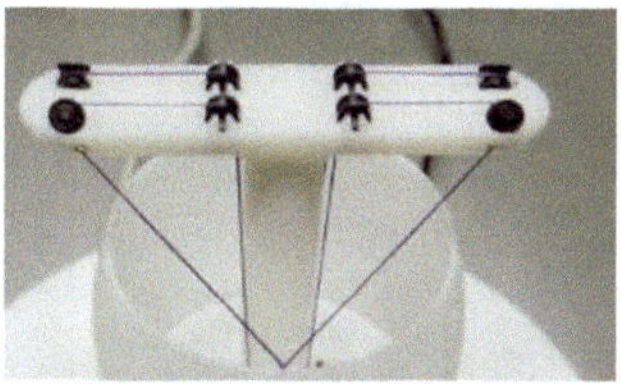

Fig. 2 Guide rollers to pull wires up

Fig. 3 New SPIDAR-I

4 Evaluation Experiments of SPIDAR-I

4.1 Measurement Experiment of Peak Force

The optimal structure for omnidirectional isotropic peak force was almost found analytically. However, to produce a haptic interface, the utilization of wire guide rollers to keep the design compact and the error of positions to infix strings makes the haptic realizing performance not necessarily performing as theoretically predicted. Therefore, we presented the force measurement experiments and compared the results with the theoretical values. The experimental devices for peak force feedback are shown in Fig. 4. The experimental device is a combination of 6-axis force sensor Dyn-Pick of Wacoh-Tech, Inc. (Model: WDF-6A50-1-UG2) and parts made by Edmund Optics, Inc. and the components are shown below.

- *XYZ*-axis Stage
  Allows ±10.0 mm of translation along *X*-axis (*Y*-axis) direction and ±20.0 mm translation along *Z*-axis direction. It is attached to the surface plate.
- Rotation stage
  Can be rotated 360°. It is attached to the *XYZ*-axis stage.
- 6-axis force sensor

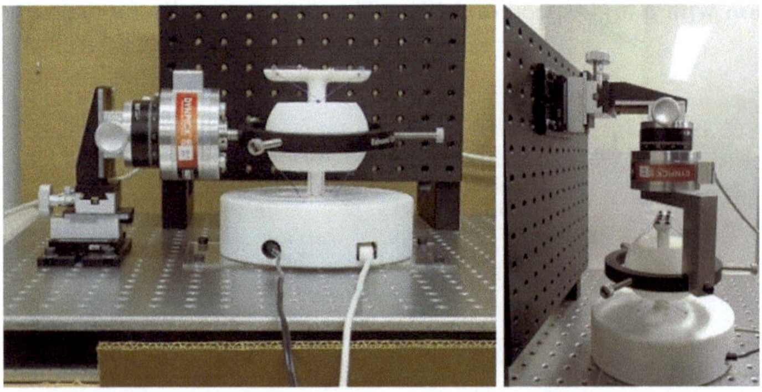

Fig. 4 Force measurement

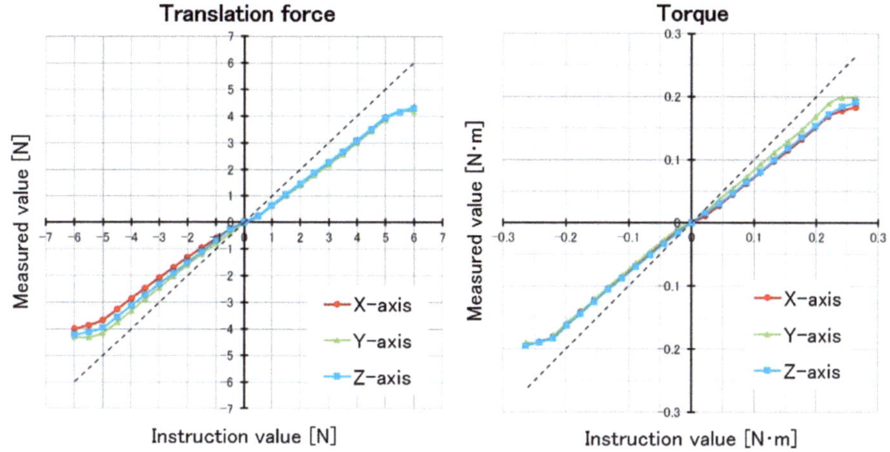

Fig. 5 Force presentation performance (*left*) and torque presentation performance (*right*)

Can measure the translational force up to 50.0 N and rotational force up to 1.0 N m along the directions of each axis. It is attached to the rotation stage.

- Grip holder

  Thing that holds the grip. It is attached to the 6-axis force sensor.

  According to the composition of experimental device described above, we measured the peak force along the directions of 3 principal axes, while the grip was in the home position. Assume that the maximum tension of each wire is 2.12 N (this value is from Sect. 3.2) and the minimum tension is 0.01 N. The measurement results of translational force and rotational force are shown in Fig. 5. The measurement results of peak force are shown in Table 2. In addition, the theoretical value of peak force from the most difficult direction to realize is 1.880 N.

  According to Table 2, although the largest difference between the theoretical value and the measured value is 0.3 N, but considering the influence of factors such as the friction of guide rollers and wires, generally we are able to conclude that the results agree with the theoretical values.

Table 2 Comparison of peak force

	Measured values		Theoretical values
	+Direction	−Direction	±Directions
Fx [N]	4.321	−4.002	±4.327
Fy [N]	4.171	−4.370	±4.434
Fz [N]	4.287	−4.223	±4.327
Mx/*R* [N]	4.164	−4.425	±4.327
My/*R* [N]	4.515	−4.307	±4.434
Mz/*R* [N]	4.338	−4.436	±4.327

Fig. 6 Results of the questionnaire

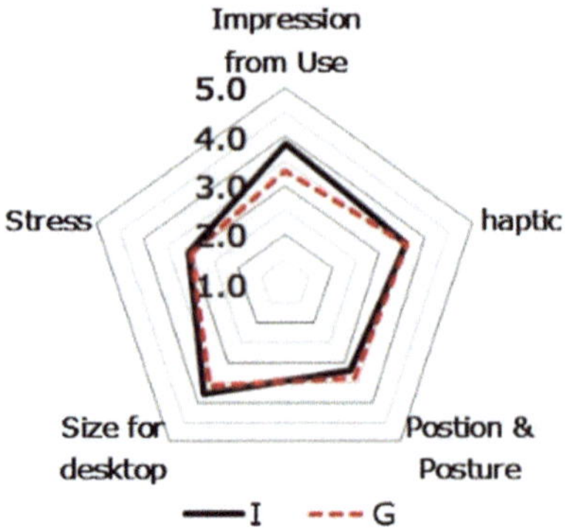

4.2 Evaluation Experiment by Task

In order to measure the performance of the haptic interface, it is necessary to actually operate the interface for evaluation. Therefore, we carried out the evaluation by performing the task of putting L-shaped block into boxes and taking questionnaire from subjects. Subjects were 10 males around 23–24 years of age.

The items of questionnaire are as follows.

- How did you feel while using the device
- Did you feel the force as expected
- Did the object moved to the position as expected
- Is the size suitable for a desktop device
- Did you feel any stress during the task

We carried out the evaluation by letting subjects grade the items from grade 1 to grade 5. The average results of the questionnaire are shown in Fig. 6. A larger grade means a better result, while a smaller grade means a worse one. The results of the questionnaire shows that, despite the small size, SPIDAR-I performed comparably to the existing device SPIDAR-G. Moreover, it is more handy than SPIDAR-G since users don't need to worry about touching wires during operation to cause an error of measuring.

5 Conclusion

In this study, we optimize the structure of the 6 DoF haptic device realized by 8 wires for omnidirectional isotropic peak force feedback, while the grip is in the home position. Then, we produced a device with the optimal structure and evaluated the device by experiments. And despite some errors, we obtained results that agree with the theoretical values in the range of measurement. Also, by performing evaluation experiment with task, it is confirmed that the proposal device has a comparable performance to the existing device as a haptic interface while being compact.

References

1. Massie, T.H.: The phantom haptic interface: a device for probing virtual objects. In: Proceedings of the ASME Dynamic Systems and Control Division, vol. 55 (1994)
2. Kawamura, S., Choe, W., Tanaka, S., Pandian, S.R.: Development of an ultrahigh speed robot FALCON using parallel wire drive systems. In: Proceedings of the IEEE International Conference on Robotics and Automation, 1995, vol. 1 (1995)
3. Hirata, Y., Sato, M.: A new interface device SPIDAR for virtual work space. In: ICIIPS'92 (1992)

Robotic Touch Surface: 3D Haptic Rendering of Virtual Geometry on Touch Surface

Seung-Chan Kim, Byung-Kil Han, Jiwon Seo and Dong-Soo Kwon

Abstract In this demonstration, we propose a robotic haptic surface display that physically imitates the orientation of virtual 3D geometry at the point of touch. The proposed system mechanically aligns its display surface with virtual 3D geometry. This allows users to obtain tactual information where contact takes place. The proposed haptic rendering scheme is built on the basic consideration that relative tactual experiences play a significant role in haptic object perception.

Keywords Surface haptics · Surface orientation · Touch surface

1 Introduction

A touch screen allows users to touch and intuitively manipulate objects displayed on the screen. Due to the two-dimensional nature of planar displays, haptic interaction with a virtual 3D object requires different access techniques. Because virtual space and real space are physically separated by the display panel, the interaction space is often limited to the display surface. This is a substantially different setup from that used for conventional haptic rendering, which generally collocates haptic information in the real interaction space [1]. As shown in Fig. 1, in this type of

S.-C. Kim (✉) · B.-K. Han · J. Seo · D.-S. Kwon
Department of Mechanical Engineering, KAIST, 291 Daehak-ro, Yuseong, Daejeon 305-701, Korea
e-mail: dalek@kaist.ac.kr
URL: http://robot.kaist.ac.kr/~kimsc/

B.-K. Han
e-mail: bkhan2000@kaist.ac.kr

J. Seo
e-mail: jwseo22@kaist.ac.kr

D.-S. Kwon
e-mail: kwonds@kaist.ac.kr

H. Kajimoto et al. (eds.), *Haptic Interaction*, Lecture Notes in Electrical Engineering 277, DOI 10.1007/978-4-431-55690-9_32

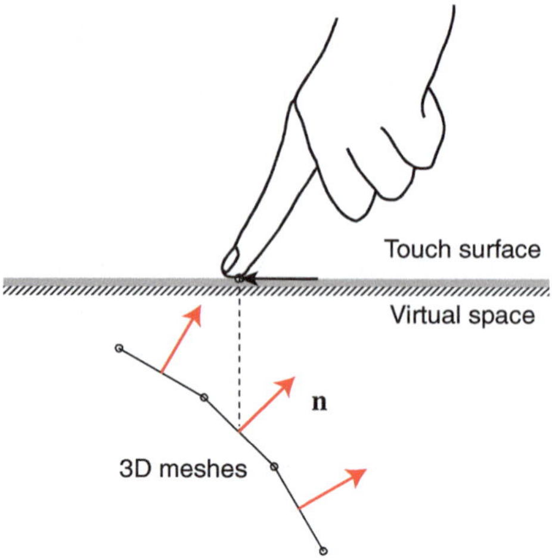

Fig. 1 Example of haptic interaction with a projected geometry

surface interaction, the HIP, e.g., the user's finger, and the proxy point cannot penetrate the rigid display surface. Because the virtual and real world are not collocated in this type of interaction, we need to redefine the contact between the HIP and virtual geometry.

2 Proposed System

2.1 Overview

One fundamental observation exploited in our system is that haptics in the real world is closely related to *relative* tactual experiences. For example, we perceive real objects based on the slope/curvature while we are touching an object. In an earlier study, Gordon and Morison revealed that the surface gradient is an effective stimulus for perceiving the object curvature regardless of the geometry of the surfaces [2]. Indeed, such a gradient feature was employed by previous haptics research, as described in Sect. 2.2 [3–6].

2.2 Hardware System

The simplest type of a commercial parallel platform with 3 linear servo motors is employed to control the posture of the display surface. The device, a RPS parallel platform, consists of an upper (moving) and a lower (fixed) platform and three prismatic joints located between the two platforms. A visual display is attached on top of the upper moving stage. The system determines the position of each actuator based on the pose of the upper platform.

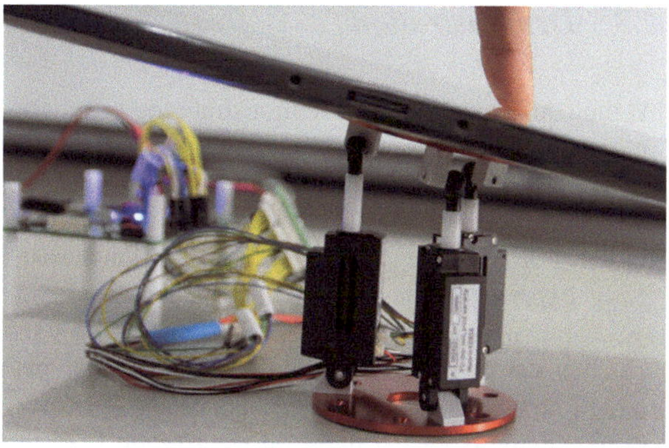

Fig. 2 Proposed robotic surface based on a simple RPS parallel platform. The surface orientation based on the touched 3D geometry would provide immersive visual and haptic experiences

We take the surface normal of virtual geometry at the touched point as an input to the system. The touched point is calculated by unprojecting window coordinates ($\in R^2$) to object coordinates ($\in R^3$) during the interaction phase. The system output is the roll and pitch of the display panel, and these variables are used to control each servo motor based on the kinematics of the platform (Fig. 2).

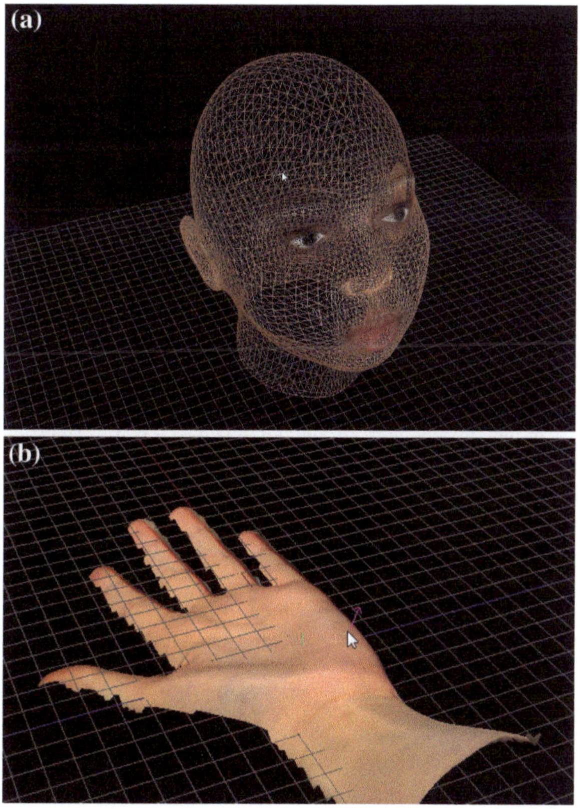

Fig. 3 Examples of applications. **a** Example of a virtual geometry. **b** Example of a captured hand

2.3 Demonstration

At the conference, we will demonstrate the proposed robotic surfaces with engaging interaction scenarios. Visitors will be able to interact with a set of interactive surfaces, each of which contains different types of 3D models ranging from generic 3D models such as virtual characters to captured 3D models such as body parts and fossils. For example, when touching virtual human head models, users will physically experience the shape of the face and head, including bumps around the eyebrows, nose, and lips. Figure 3b shows an example of a social application in which users can physically interact with other people through the captured data model. In this application, users are able to physically feel the bumps and grooves on the captured virtual hand on the robotic screen.

Acknowledgements This work was supported by the ICT R&D program of MSIP/IITP. [4-000-11-001, Human Friendly Devices (Skin Patch, Multi-modal Surface) and Device Social Framework Technology].

We would like to thank Dr. Ivan Poupyrev and Dr. Ali Israr for the inspiration of the gradient-based approach in the context of surface haptics.

References

1. Kim, S.-C.: Haptic interaction with 3D geometric objects on flat touch surfaces: gradient-based approach. Ph.D. Dissertation, Department of Mechanical Engineering, KAIST (Korea Advanced Institute of Science and Technology) (2013)
2. Gordon, I.E., Morison, V.: The haptic perception of curvature. Atten. Percept. Psychophys. **31**, 446–450 (1982)
3. Minsky, M., Ming, O.-Y., Steele, O., Frederick, J., Brooks, P., Behensky, M.: Feeling and seeing: issues in force display. SIGGRAPH Comput. Graph. **24**, 235–241 (1990)
4. Robles-De-La-Torre, G., Hayward, V.: Force can overcome object geometry in the perception of shape through active touch. Nature **412**, 445–448 (2001)
5. Saga, S., Deguchi, K.: Lateral-force-based 2.5-dimensional tactile display for touch screen. In: IEEE Haptics Symposium, pp. 15–22 (2012)
6. Kim, S.-C., Israr, A., Poupyrev, I.: Tactile rendering of 3D features on touch surfaces. Presented at the proceedings of the 26th annual ACM symposium on user interface software and technology, St. Andrews, Scotland, UK (2013)

SRU: Stepwise Rotation Update of Finite Element Model for Large Deformation

Yoshihiro Kuroda and Haruo Takemura

Abstract A deformable model has been studied for applications in education, entertainment, and medicine. Large deformation of a linear finite element model induces undesirable volume expansion, due to nonlinearity of element's rotation. We demonstrate stepwise rotation update of largely rotated elements, which improves a quality of deformation and reduces computation at a step effectively. A user can manipulate an object with force feedback, although conventional nonlinear finite element methods require offline simulation.

Keywords Finite element method · Large deformation · Haptics · Nonlinearity

1 Introduction

Physics-based deformable models have been studied intensively in computer graphics and computational mechanics. However, the computational cost is one of the main issues for interactive system, such as a video game, training, and navigation systems.

Y. Kuroda (✉) · H. Takemura
Cybermedia Center, Osaka University, 1-32 Machikaneyama-cho, Toyonaka, Osaka 560-0043, Japan
e-mail: ykuroda@ime.cmc.osaka-u.ac.jp

Y. Kuroda · H. Takemura
Graduate School of Information Science and Technology, Osaka University, 1-5 Yamadaoka, Suita, Osaka 565-0871, Japan

H. Kajimoto et al. (eds.), *Haptic Interaction*, Lecture Notes in Electrical Engineering 277, DOI 10.1007/978-4-431-55690-9_33

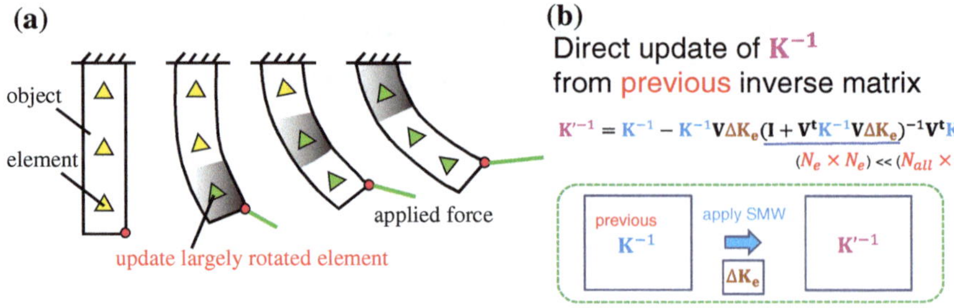

Fig. 1 Concept of stepwise rotation update: **a** stepwise update of element stiffness matrix of largely rotated elements, **b** direct update of inverse stiffness matrix

2 Related Works and Our Approach

Finite element model is a major solution to represent physical behavior of the elastic material based on continuum mechanics. Hirota et al. proposed a method to calculate reaction force from stiffness equation by precalculating inverse of stiffness matrix. It enables to calculate plausible reaction force in real time, when the deformation is small enough [1]. Müller proposed a corotation method to change stiffness matrix of an element for reducing deformation artifact in large deformation [2]. Some related works proposed methods to calculate deformation [3, 4]. However, the above methods require much computation to calculate force from stiffness equation. In this demonstration, we implemented a stepwise rotation update (SRU) method for interactive manipulation of a FEM-based deformable object, which was proposed in [5, 6].

Figure 1 shows a concept of the implemented system. The system equips with a haptic interface, PHANToM by SensAble. A user can manipulate an object interactively and experience the difference of the performance of the existing models and the SRU method.

Acknowledgments This work was partly supported by Grant-in-Aid for Scientific Research (B) (26282147).

References

1. Hirota, K., Kaneko, T.: Haptic representation of elastic objects. MIT Presence Teleoper. Virtual Environ. **10**(5), 525–536 (2001)
2. Müller, M., Gross, M.: Interactive virtual materials. Proc. Graphics Interface **2004**, 239–246 (2004)
3. Kikuuwe, R., Tabuchi, H., Yamamoto, M.: An edge-based computationally efficient formulation of Saint Venant-Kirchhoff tetrahedral finite elements. ACM Trans. Graphics **28**(1), 1–13 (2009)

4. Sin, F., Schroeder, D., Barbic, J.: Vega: non-linear FEM deformable object simulator. Comput. Graph. Forum **32**(1), 36–48 (2013)
5. Kuroda, Y., Uranishi, Y., Imura, M., Oshiro, O.: Finite element simulation of large deformation for haptic interaction. Trans. VRSJ **18**(4), 497–506 (2013) (in Japanese)
6. Kuroda, Y., Uranishi, Y., Imura, M., Oshiro, O., Takemura, H.: Large Deformation with Haptic Interaction by Stepwise Rotation Update of Finite Element Model. Prof. of CARS, vol. 9, Suppl.1, S127-128, pp. 497–506 (2014)

A Conceptual Design of a Smart Knob with Torque Feedback for Mobile Applications

Sang Kyu Byeon, Dong-Soo Choi, Won-Hyeong Park, Yu-Joon Kim, Ki-Uk Kyung and Sang-Youn Kim

Abstract This study proposes a conceptual design of a tiny smart knob based on MR fluids in order to convey realistic haptic feedback to users in mobile devices. The haptic sensation, which is created in the form of resistive torque, is varied according to the current input. The proposed knob is designed in small size as possible as can so that it is easily inserted into mobile devices.

Keywords Torque feedback · Haptic sensation · Mobile device · Game · Entertainment

1 Introduction

Although tactile information is widely used for manipulating with mobile devices, the kinesthetic information is equally important in addition to the tactile information. The kinesthetic sense can be generated by the stiffness of the object, and conveying the degree of stiffness is a key factor to precisely feel the objects. However, there have been limited research works for developing stiffness varying modules. Fujita and Ohmori [1] developed a stiffness display by regulating a pump for fluid volume control. Liu et al. [2] presented a stiffness display consisting of a thin elastic beam and an actuator to adjust the deformable length of the beam. Badshah et al. [3] suggested a multi-DOF (degree of freedom) torque feedback system using gyro effect. Winfree et al. [4] developed an ungrounded haptic device which creates torque and conveys it to a user.

S.K. Byeon · D.-S. Choi · W.-H. Park · Y.-J. Kim · S.-Y. Kim (✉)
KoreaTech, Chungjello 1600, Byeongcheon-Myeon, Cheonan, ChungNam Province 330-708, South Korea
e-mail: sykim@kut.ac.kr

K.-U. Kyung
ETRI, Gajeongro 210, Yusenggu, Daejeon 305-700, South Korea

H. Kajimoto et al. (eds.), *Haptic Interaction*, Lecture Notes in Electrical Engineering 277, DOI 10.1007/978-4-431-55690-9_34

Even though these systems provide large enough force or torque feedback to stimulate a human hand in small-sized device, their size is rather bulky to be embedded into a mobile device. In order to create kinesthetic button sensation in mobile devices without any stability problems, small stiffness display has been developed using MR fluid [5].

There are some buttons or rotary knobs for providing a user's command to mobile devices or electronic devices. Whenever a user manipulates a target device using a rotary knob, he/she wants to receive torque feedback. Therefore, this study suggests a smart knob which conveys torque feedback to a user. For developing the smart knob, we use MR fluids which allow the proposed knob to have a simple mechanical structure.

2 The Proposed Smart Knob

Figure 1a shows the overall structure of the proposed smart knob consisting of three parts: (1) an interacting part, (2) a torque tuning part, and (3) a flux creating part. The interacting part includes a contactor and a silicon O-ring for preventing MR fluids leak. The flux creating part, which acts as the magnetic path minimizing the loss of magnetic flux, consists of a housing cover and housing. The torque tuning part consists of MR fluids, a solenoid coil, a spacer, a linear guide, and a plunger. The solenoid creates magnetic fields to control the stiffness of MR fluids which is filled in the housing. Figure 1b shows the fabricated smart knob. If the electric current is applied to the solenoid coil in order to generate magnetic field, particle chains are built up in the knob. Therefore, according to the amount of input current, the strength of the feedback torque can be changeable.

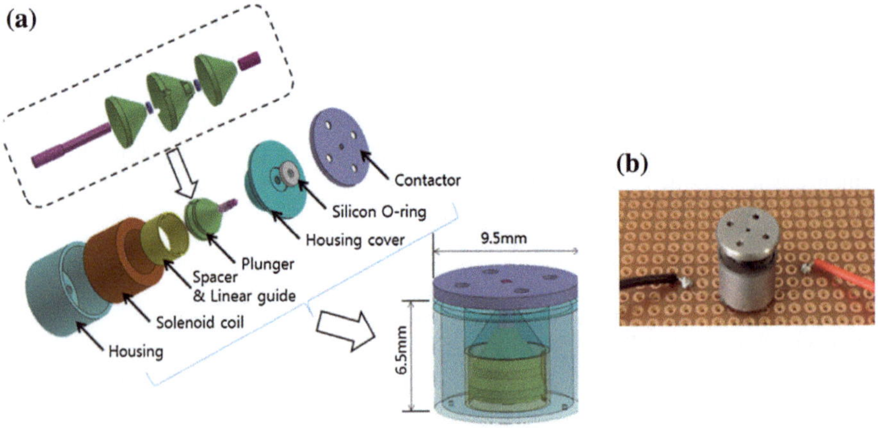

Fig. 1 Schematic illustration of the proposed actuator (**a**) and the fabricated smart knob (**b**)

3 Conclusion

This study has presented a new smart knob which creates torque sensation. The proposed knob consists of three major parts, such as an interacting part, a torque tuning part, and a flux creating part. Once the electric current is applied to the solenoid coil, MR fluids build up particle chains in the knob so that a user can sense torque sensation.

Acknowledgements This research was supported by the Pioneer Research Center Program through the National Research Foundation of Korea funded by the Ministry of Science, ICT and Future Planning (NRF-2013M3C1A3059588).

References

1. Fujita, K., Ohmori, H.: A new softness display interface by dynamic fingertip contact area control. In: 5th World Multiconference on Systemics, Cybernetics and Informatics, pp. 78–82 (2001)
2. Liu, J., Song, A., Zhang, H.: Research on stiffness display perception of virtual soft object. In: International Conference on Information Acquisition (2007)
3. Badshah, A., Gupta, S., Morris, D., Patel, S.N., Tan, D.: GyroTab: a handheld device that provides reactive torque feedback. In: SIGCHI Conference on Human Factors in Computing Systems, pp. 3153–3156 (2012)
4. Winfree, K.N., Gewirtz, J., Mather, T., Fiene, J., Kuchenbecker, K.J.: A high fidelity ungrounded torque feedback device: the iTorqU 2.0. In: IEEE World Haptics Conference, pp. 261–266 (2009)
5. Yang, T.H., Kwon, H.J., Lee, S.S., An, J., Koo, J.H., Kim, S.Y., Kwon, D.S.: Development of a miniature tunable stiffness display using MR fluid for haptic application. In: Sensors and Actuators A: Physical, A. 163, pp. 180–190 (2010)

Part IV
Sensing

Built-in Capacitive Position Sensing for Multi-user Electrostatic Visuo-haptic Display

Taku Nakamura and Akio Yamamoto

Abstract This article describes a built-in capacitive-type position sensing for a multi-finger visuo-haptic display system using electrostatic force. The sensing system shares the same components with the electrostatic haptic rendering, which are voltage signal sources, contact pads, and a sheet of ITO electrode. High-frequency sensor signal is superposed onto low-frequency high-voltage signal for the electrostatic haptic feedback, and the flowing high-frequency current is measured to estimate the position of the pad. To demonstrate the system capability, hockey game was implemented, in which two players can hit a virtual puck with haptic feedback.

Keywords Surface haptics · Passive haptics · Surface capacitive touch screen · Haptic rendering

1 Introduction

Recently, haptic/tactile feedback on a large-size visual display has been a hot topic [1]. To realize multi-finger/multi-user haptic feedback on a large-size display, we have developed several visuo-haptic systems using electrostatic friction modulation [2–4]. The system detects the position of fingers and provides appropriate feedback using electrostatic force. The position detection in those systems was an overhead camera (or Wii-mote) [2, 3], or an infrared-vision-based touch panel [4], which suffer from occlusion or low processing speed, respectively. To overcome the problems, this work proposes a built-in capacitive sensing system for the multi-user electrostatic visuo-haptic system.

T. Nakamura (✉) · A. Yamamoto
The University of Tokyo, 7-3-1, Hongo, Bunkyo-ku, Tokyo 113-8656, Japan
e-mail: taku_nkat@aml.t.u-tokyo.ac.jp

A. Yamamoto
e-mail: akio@aml.t.u-tokyo.ac.jp

H. Kajimoto et al. (eds.), *Haptic Interaction*, Lecture Notes in Electrical Engineering 277, DOI 10.1007/978-4-431-55690-9_35

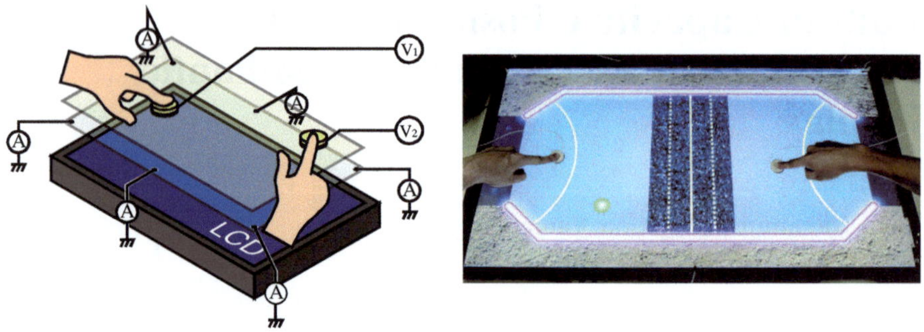

Fig. 1 Schematic illustration of prototype setup and appearance of application

2 Setup

Figure 1 (left) schematically illustrates the proposed system. The system consists of two voltage sources, two contact pads, a sheet of surface-insulated ITO electrode that covers the surface of the display, and current detection circuits. By applying high-voltage ac (up to 500 $V_{0\text{-}p}$, 400 Hz), electrostatic attraction force is generated between the pad and the display surface. As the user moves the pad, the attraction force is converted into friction, which functions as passive haptic feedback force. The sensing is carried out by superposing a high-frequency (100 kHz) sensing signal to the haptic signal. To facilitate the measurement for two pads, the sensing signal voltage is alternatively superposed to the two pads. To reduce the interference between two pads, the location of the pads was restricted to left- and right-hand sides of the display, respectively. The position of each pad was estimated from the amplitudes of the high-frequency currents measured at the four corners of each half, using predefined calibration maps. The linearity after the calibration was a few percent of the full scale in this particular work. The sensor and haptic rendering loop ran at 200 Hz, whereas visual rendering was performed at 60 Hz.

Each voltage source mainly consists of a high-voltage amplifier and a transformer for sensor signal superposition. A current limiting resistor (150 kΩ) was connected to the output of the high-voltage amplifier for safety reasons [2].

3 Application: Hockey Game

A hockey game was implemented to demonstrate the proposed system. Figure 1 (right) shows the appearance of the application. In this application, two players can interact with the virtual puck and the hockey court through the contact pads. When the user hits the puck, an impulsive friction force is fed back to the user. When the user hits the virtual wall with the pad, the system feeds back large friction to impede penetration. The users can also feel surface textures of the dart in the middle of the court.

Acknowledgement This work was supported by Grant-in-Aid for Scientific Research (B) (No. 26280069) and Grant-in-Aid for JSPS Fellows (No. 269272) from JSPS, Japan.

References

1. Sinclair, M., et al.: TouchMover 2.0-3D touchscreen with force feedback and haptic texture. In: Proceedings of 2014 IEEE Haptics Symposium, pp. 1–6 (2014)
2. Nakamura, T., Yamamoto, A.: Multi-finger electrostatic passive haptic feedback on a visual display. In: Proceedings of IEEE WHC 2013, pp. 37–42 (2013)
3. Nakamura, T., Yamamoto, A.: Multi-finger surface visuo-haptic rendering using electrostatic stimulation with force-direction sensing gloves. In: Proceedings of 2014 IEEE Haptic symposium, pp. 489–491 (2014)
4. Nakamura, T., Yamamoto, A.: Position and force-direction detection for multi-finger electrostatic haptic system using a vision-based touch panel. In: Proceedings of IARIA ACHI 2014, pp. 160–165 (2014)

Highly Flexible and Transparent Skin-like Tactile Sensor

Saekwang Nam, Suntak Park, Sungryul Yun, Bong Je Park, Seung Koo Park, Mijeong Choi and Ki-Uk Kyung

Abstract This demonstration shows a highly flexible, transparent, and thin tactile sensor. The sensor detects touch forces at single or multiple points with fast response and high bendability in response to dynamic input force (0–3 N) without any electronic components on sensing areas. The sensor is also capable of sensing pressure on curvilinear soft surface such as human skin without significant performance degradation. The force sensor has potentials to be used in a number of applications for measuring dynamic contact forces on various surfaces.

Keywords Flexible · Transparent · Tactile · Sensor · Wearable

1 Sensor Design

Our flexible tactile sensor is prepared by integrating a thin touch layer onto a polymer waveguide structure. The polymer waveguide is fabricated by multi-step lithography process using two pre-polymers with different refractive index [1, 2]. Theoretically, different from a KINOTEX sensor [3], the waveguide is composed of clad and core. The clad layer fully covers the core to prevent the light from scattering at undesired areas during the propagation of the light through the waveguide. Only at the sensing areas with bare core, the light can be easily scattered from the core. The sensor, which is designed to an array of 12 sensing points at 4 × 3 matrix similarly to a numeric keypad, allows for detection of touch forces at each localized areas with a size of 5 mm (width) × 5 mm (length). Since the sensor works without any electrical components in sensing areas, it is capable of detecting the touch forces under a bending condition attaching the sensor to soft and flexible interface such as forearm (Fig. 1). The benefit also enables the sensor to operate in water.

S. Nam · S. Park · S. Yun · B.J. Park · S.K. Park · M. Choi · K.-U. Kyung (✉)
Transparent Transducer and UX Creative Research Center, ETRI, Daejeon, South Korea
e-mail: kyungku@etri.re.kr

H. Kajimoto et al. (eds.), *Haptic Interaction*, Lecture Notes in Electrical Engineering 277, DOI 10.1007/978-4-431-55690-9_36

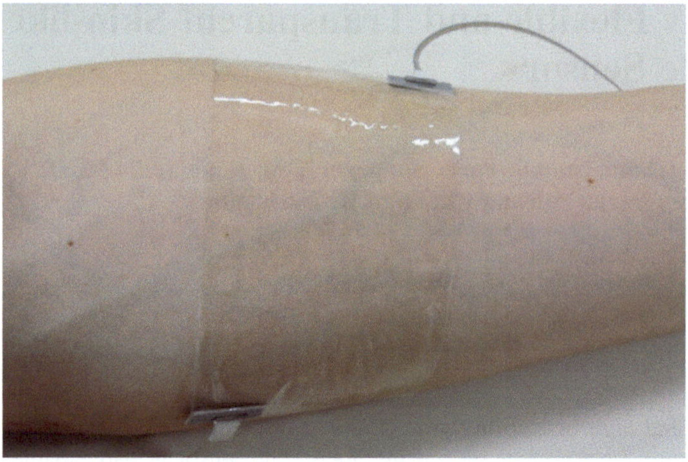

Fig. 1 A photography of the sensor attached to a forearm

2 Working Principle

The sensor consists of the polymer waveguide and a thin touch layer. The polymer waveguide is connected to a light source using a laser diode and a photodetector. Without any pressurized contact of the touch layer with a bare core in each sensing area, there is no significant light loss during propagating the light from the source through the waveguide due to the use of core with relatively higher refractive index than clad. On the other hand, when the touch layer is in contact with the bare core at a sensing area in response to a pressured input, the light passing through the core can be scattered toward the surface of touch layer. Since the amount of scattering of light passing through the core can be changed with contact area between the touch layer and the core, our force sensor detects touch pressure in the range of 0 and 3 N by monitoring output light intensity (L_i) with respect to input L_i from a source using photodetectors, which are independently connected to each core. Figure 2 shows a working principle of our sensor.

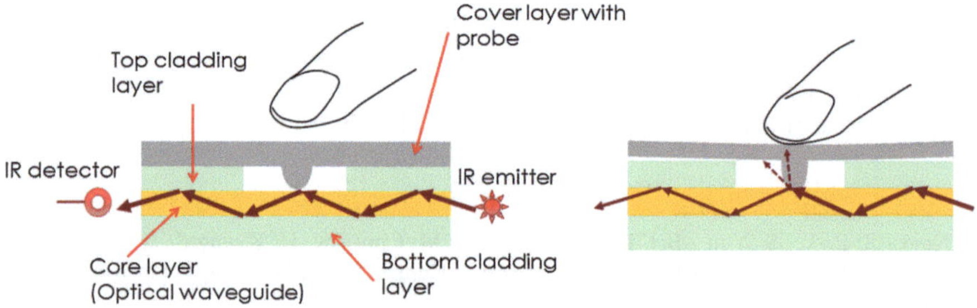

Fig. 2 A working principle of the transparent and flexible sensor: a part of infrared lights leaks from the core layer at the sensing area when the touch layer is contacting with the core

3 Applications

Our polymer waveguide-based sensor is capable of working on curvilinear interfaces such as flexible displays or human skin. Therefore, this benefit allows the sensor to be adapted for future wearable device such as a force detectable sensor on a watchstrap or a biological sensor checking any perceived force on a body.

Acknowledgements This work was supported by the ETRI and MKE/KEIT, [Development of Transparent Actuator and UX, TAXEL: Visio-haptic Display and Rendering Engine (10035360)].

References

1. Yun, S., Park, S., Park, B., Kim, Y., Park, S.K., Nam, S., Kyung, K.U.: Polymer-waveguide-based flexible tactile sensor array for dynamic response. Adv. Mater. **26**, 4474–4480 (2014)
2. Kim, Y., Park, S., Park, S.K., Yun, S., Kyung, K.U., Sun, K.: Transparent and flexible force sensor array based on optical waveguide. Opt. Express **20**(13), 14486–14493 (2012)
3. Sakai, K., Nakagami, G., Matsui, N., Sanada, H., Kitagawa, A., Tadaka, E., Sugama, J.: Validation and determination of the sensing area of the KINOTEX sensor as part of development of a new mattress with an interface pressure-sensing system. BioScience Trends **2** (1), 36–43 (2008)

A Mounting Foot-Type Force-Sensing Device for a Desk with Haptic Sensing Capability

Toshiaki Tsuji, Tatsuki Seki and Sho Sakaino

Abstract This study deals with the development of a force-sensing device for converting a desk into an interface by mounting the device on the supports or legs of these objects. The system is composed of four force sensors with clamps, and a PC. Contact force and its position on a regular desk can be estimated from the response of the four force sensors fixed on the legs of the desk. Some application programs using this interface are introduced.

Keywords Force sensing · Tactile sensing · Haptic sensing · Interface · Touch panel

1 Introduction

This research proposes a technology for converting a desk into an interface by installing force sensors on their legs. Since long, many studies have focused on the principles of identifying tactile sensing information by using force sensors. Tactile-sensing information, in the context of this study, refers to the locations of force application points and their vectors. Bicchi [1] indicated that this theory can be extended to elastic body links. Iwata et al. [2] have developed a whole-body-sensing interface by applying such theories to human-assisting robots. The interface developed by Iwata et al. offers an advantage of accurately detecting the sum of external forces acting on the link, but there is still room for improvement in terms of reducing the amount of wiring as well as sensors since additional touch sensors are used to determine the contact surfaces [3]. The authors developed the whole-body-type tactile sensor, "Haptic Armor," and demonstrated that sensors and wiring are

T. Tsuji (✉) · T. Seki · S. Sakaino
Graduate School, Saitama University, 255, Shimo-okubo, Sakura, Saitama, Japan
e-mail: tsuji@mail.saitama-u.ac.jp

S. Sakaino
e-mail: sakaino@mail.saitama-u.ac.jp

H. Kajimoto et al. (eds.), *Haptic Interaction*, Lecture Notes
in Electrical Engineering 277, DOI 10.1007/978-4-431-55690-9_37

not required on the outer shell to obtain the tactile information with force sensors alone, even when the external shell shapes were discontinuous [4]. This theory was extended to non-convex shape [5] and soft material. These techniques are applied as a desk equipped with force sensors, the "Haptic Desk" [6,7]. FuwaFuwa[8]is an interface with similar effects, but it is constrained by the requirement that the interface material must be pliable. Furthermore, studies have also focused on the conversion of a desktop into an interface like a canvas, by using a camera to detect a finger [9],but such systems were not operable unless the scope is maintained within the viewing angle of the camera; further, the entire desktop could not be used as an interface. This chapter describes the development of force-sensing devices for converting a regular desk into an interface. This device is composed of four force sensors and a PC to process the sensing information. The structure of the device is such that they could be fitted on supports or legs of desks.

2 Configuration of Developed System

2.1 Mechanism

The application of the haptic armor theory [7] makes it possible to detect the external force application points even without installing tactile sensors on the surface skin of objects. In the mechanism developed in this research, force sensors are installed on the supports of objects subject to conversion into interfaces. A mechanism with a fixture attached to a force sensor was developed. The setup is shown in Fig. 1. The end effector of the force sensor was secured to a leg using a clamp mechanism tightened by rotating the grip. The six-axis force sensor, CFS034CA301U from Leptrino, was used as the force sensor. These sensors could be replaced by other three-axis force sensors, since force information of only three axes was used to calculate the contact position.

Fig. 1 The force-sensing device with the mounting mechanism

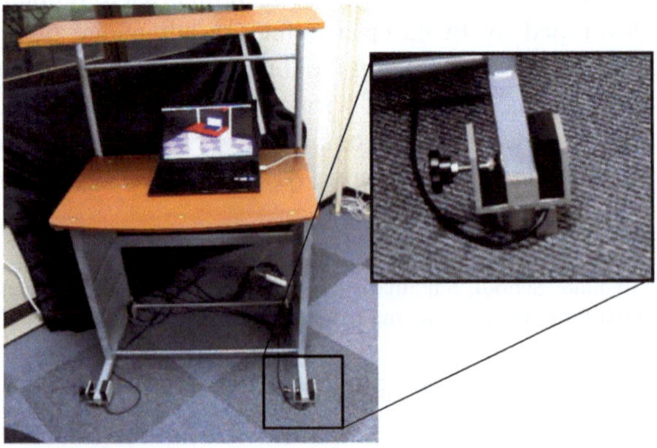

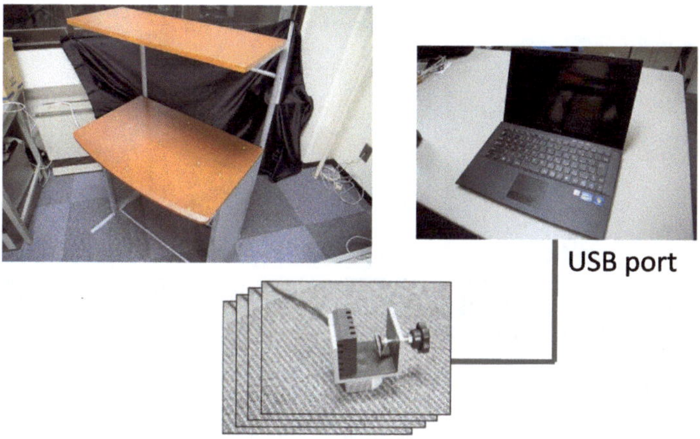

Fig. 2 System configuration

2.2 System Configuration

The actual system configuration is shown in Fig. 2. The objects that were converted into interfaces were not included in the basic configuration, since this research was an attempt to convert a regular desk into an interface. This system comprised force sensors installed on the legs of objects as well as a USB hub and a laptop PC. The theory of this chapter can be applied so long as the force and the resultant force of moments of the six axes can be derived, even if the number of sensors is decreased. The information from the force sensors is sent to the laptop PC, and the external forces as well as their application points are identified by a software process running on the PC. The objects can be used as interfaces for the PC, by sending the information to the mouse driver or software.

2.3 Entry of Geometrical Shapes of Interfaces

The basic principle of calculating contact location is shown in [3]. In addition to the response values obtained from the force sensors, the geometrical shape of the object should be known for the algorithm discussed in this research. Generally, inputting the geometrical shape is a tedious task; however, if only objects with simple shapes, such as a desk, are considered, and if a model is prepared in advance, then the geometrical shape can be input by simply entering the dimensions. Hence, in this study, a simple rectangular desk was selected as the object, and its model was input into the program in advance. A variety of desks can be considered simply by entering the three parameters: length, width, and height.

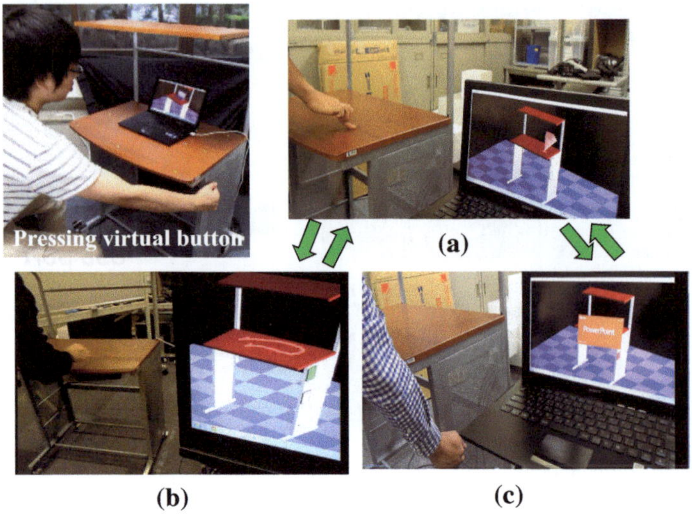

Fig. 3 Three modes switched by virtual button. **a** Basic mode. **b** Drawing mode. **c** Power point mode

2.4 Application Program of Haptic Desk

As shown in Fig. 3, there are three modes on the application program of the haptic desk: (a) basic mode, (b) drawing mode, and (c) powerpoint mode. Additionally, object recognition based on loads can be performed, regardless of the currently executed mode. (a) The basic mode displays the contact points and external force vectors. (b) The drawing mode allows the continuous drawing of contact points and the retention of these drawing strokes. The thickness of the drawing strokes can be controlled according to the sizes of the external force vectors. (c) In the powerpoint mode, powerpoint is manipulated by pressing virtual switches. Operating the switches on the side frame makes it possible to switch from basic to other modes, delete drawing stroke history, and select the surfaces of end effectors in use. The switches on the top board can be used while in the powerpoint mode as well. Six types of operations are set with respective switches: "→" "←," "Esc," "F5," and "Shift + F5." Switches on the top board and the side frame were located within a square domain with sides of 0.1 m.

3 Conclusion

This chapter described the development of a force-sensing device for converting a desk into an interface by mounting the device on the supports or legs of these objects. A variety of shapes must be acceptable to the program in order to assume that the device is to be mounted on regular desks. If a geometric model of the desk is given in advance, the external forces and their application points can be identified by simply entering their dimensions. Some application programs utilizing the property of the proposed interface are developed. The activated programs are selected by the virtual buttons created on the regular desk.

Acknowledgements We would like to mention that a portion of this research was funded by the Grant-in-Aid for Scientific Research Program provided by the Japan Society for the Promotion of Science (Topic No. 24560539). We would also like to mention that we received cooperation from Mr. Shota Ito and Mr. Takeshi Kaneko of the Graduate School of this university in building the experiment equipment used for this paper.

References

1. Bicchi, A.: Intrinsic contact sensing for soft fingers. In: Proceedings of the IEEE International Conference Robotics and Automation, pp. 968–973 (1990)
2. Iwata, H., Hoshino, H., Morita, T., Sugano, S.: The whole-body sensing interface for robots that coexist with humans. J. Robot. Soc. Jpn. **20**(5), 543–549 (2002)
3. Tsuji, T., Kaneko, Y., Abe, S.: Whole-body force sensation by robot with outer shell. In: Proceedings of 10th IEEE International Workshop on Advanced Motion Control (AMC '08), pp. 365–370 (2008)
4. Tsuji, T., Kaneko, Y., Abe, S.: Whole-body force sensation by force sensor with shell-shaped end-effector. IEEE Trans. Ind. Electron. **56**(5), 1375–1382 (2009)
5. Kurita, N., Sakaino, S., Tsuji, T.: Whole-body force sensation by force sensor with end-effector of arbitrary shape. In: Proceedings of IEEE/RSJ International Conference on Intelligent Robots and Systems, pp. 5428–5433 (2012)
6. Tsuji, T., Kurita, N.: Development of interface using tactile technology by force sensor. In: The Robotics and Mechatronics Conference 2014 in Toyama, 1P1-X05, The Japan Society of Mechanical Engineers (2014)
7. Sugiura, Y., Kakehi, G., Withana, A., Lee, C., Sakamoto, M., Sugimoto, M., Inami, M., Igarashi, T.: Detecting shape deformation of soft objects using directional photoreflectivity measurement. In: Proceedings of Annual ACM Symposium on User Interface Software and Technology, pp. 509–516 (2011)
8. Sato, Y., Oka, K., Koike, H.: Video-based tracking of user's motion for augmented desk interface. In: Proceedings of IEEE International Conference on Automatic Face and Gesture Recognition, pp. 805–809 (2004)
9. Kurita, N., Hasunuma, H., Sakaino, S., Tsuji, T.: Simplified whole-body tactile sensing system using soft material at contact areas. In: Proceedings of IEEE International Conference on Industrial Electronics Society, pp. 4264–4269 (2013)

Thumbnail Input for Head-Mounted Display

Yasutoshi Makino

Abstract In this work, we propose an input interface for mobile head-mounted display. A camera is attached on frames of glasses to take pictures below. When users lift their hand so that the camera can see their thumbnail, the system recognizes it and detects its position. Measured position is used for inputting command for controlling the head-mounted display. The system also recognizes pinching force between a thumb and an index finger. The color of the thumb changes depending on the applied force. Users can interact with devices with small motions.

Keywords Haptic interface · Input device · Head-mounted display

1 Introduction

In this work, we propose a simple input interface for a mobile head-mounted display, such as the GoogleGlass. In general, people use an input interface by following 3 steps: (1) input information, (2) check the information whether it is OK, and (3) decide it. For instance, when a user clicks an icon on a screen with a mouse, (1) move the mouse, (2) check that the cursor is on the target icon, and then (3) click it. Important thing is that the second step always requires visual or auditory feedback. General haptic-based input interfaces are usually used for the first and third steps but cannot be used as the tool for the second step. Thus, the system has to have visual/auditory feedback equipment for that purpose. There have been some researches which feedbacks haptic information for the second step; however, they are difficult to use since haptic is not sensitive enough to recognize several different symbols. As a result, almost all haptic input interfaces have visual display which makes the dimension of the whole system large.

Y. Makino (✉)
The University of Tokyo, 5-1-5 Kashiwanoha, Kashiwa-shi, Chiba-ken 277-8561, Japan
e-mail: yasutoshi_makino@k.u-tokyo.ac.jp
URL: http://www.hapis.k.u-tokyo.ac.jp/public/makino/

H. Kajimoto et al. (eds.), *Haptic Interaction*, Lecture Notes
in Electrical Engineering 277, DOI 10.1007/978-4-431-55690-9_38

In this work, we assume the situation that people wears mobile head-mounted display. Thus, we do not need to worry about the dimensions of the visual feedback equipment for the second step. When people wear a visual display in their daily lives, the input interface itself can be small and simple. User can input simple command with a small motion of their finger, for example. In previous studies, some research groups proposed small input interfaces which detected subtle finger motions with magnetic sensors attached on the fingertip [1, 2]. Both researches achieved precise cursor control. Our system does not need to put any equipment on user's finger.

2 Prototype System

Figure 1 shows our system overview. We use a camera put on the frames of the glasses. The camera recognizes user's thumbnail and detects its position for inputting 2-D location. In this case, users do not need to put any devices on their fingers. We created a Haar-like features classifier for recognizing thumbnail from the image with the OpenCV. This type of classifier is usually used for face recognition.

Figure 2 shows the process flow of our prototype system. In the first phase, the system searches fingernail from the whole image. Then, we cut out the region whose width and height was as twice as larger than that of fingernail as shown in Fig. 3. In the second step, we only searched the fingernail in this small image. In this case, it took only 0.038 ± 0.0069 s for thumbnail detection. This is about 26 fps, which is similar to the refresh rate of camera. After detecting the fingernail, we used the x, y coordinates in the camera for representing position.

Brightness of the thumbnail changes depending on the pinching force between a thumb and an index finger. The basic idea to detect color change of the fingernail to

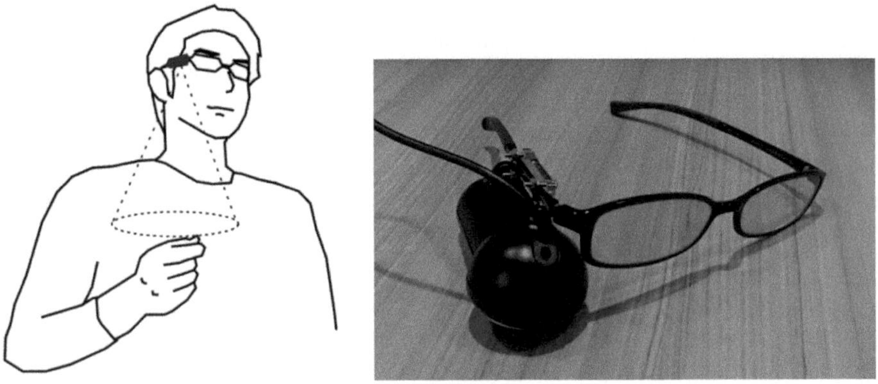

Fig. 1 System overview. Web camera is set on the frame of the glasses

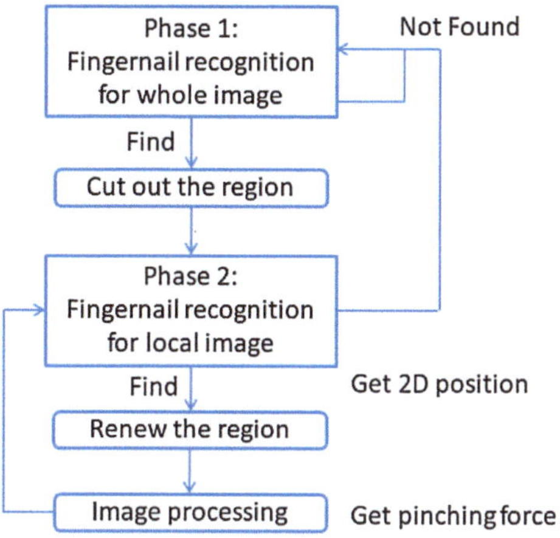

Fig. 2 Process flow of the prototype

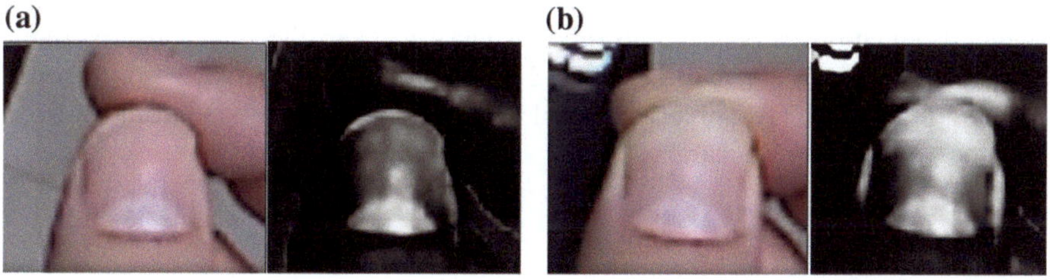

Fig. 3 Brightness of the thumbnail changes depending on the pinching force. **a** No pinching force. **b** With pinching force

estimate given force has been proposed in previous works [3]. We used this method in our system.

In order to detect the pinching force between a thumb and an index finger, a taken image was processed. At first, the image was converted into three RGB images. Then, we chose the Green image for processing since it seemed better to detect bright area of the fingernail. The contrast and the gamma value of the image were changed so that we can see the change of the fingernail color easily.

Figure 3 shows the original image of the fingernail and its processed image. The parameters for processing were determined experimentally. The averaged brightness of the tip of the fingernail relates to the applied force between a thumb and an index finger. Figure 3b shows the images when the tip of the thumb was pushed by the index finger as if they tightly pinched something thin. The brightness of the tip of the thumb differs from Fig. 3a. As a result, we can know the applied force by averaging the brightness values.

Table 1 Measured position at 5 random positions

Target position		Measured position for 8 s	
x	y	x	y
200	200	200.7 ± 2.59	199.4 ± 3.91
300	200	293.5 ± 4.45	193.2 ± 5.11
300	300	299.7 ± 2.35	300.2 ± 3.53
400	100	401.2 ± 4.85	105.4 ± 5.40
500	100	501.8 ± 6.42	105.9 ± 4.95

3 Basic Performance

We did experiment to know a basic performance of our prototype: How accurate users can control a cursor on the screen. The experiment was done by following procedures. At first on the screen, the target location was displayed as a colored circle image. One subject was asked to move the finger to reach the target area and keep it so that the cursor stays at the target location for more than 8 s. We chose 5 positions shown in Table 1 in random for the target locations.

The result was shown in Table 1. They show averaged measured position and its standard deviation. It was not so difficult tasks to keep the finger at a particular position. Maximum deviation was 6.42 pixels. If we assume that the fingernail width in the camera (100 pixels) is equal to 2 cm, then users can control their finger position with 1.2 mm accuracy.

4 Conclusions

We proposed a new input interface especially for mobile head-mounted display. The camera attached on the frames of the glasses detects fingernail. Its position and color can be used for input interface. With our device, users can control the head-mounted device without being noticed by others since the system requires only small motions.

References

1. Chen, K.Y., Lyons, K., White, S., Patel, S.: uTrack: 3D input using two magnetic sensors. In: UIST'13, pp. 237–244 (2013)
2. Chan, L., Liang, R.H., Tsai, M.C., Cheng, K.Y., Su, C.H., Chen, M.Y., Cheng, W.H., Chen, B. Y.: FingerPad: private and subtle interaction using fingertips. In: UIST'13, pp. 255–260 (2013)
3. Mascaro, S., Asada, H.: Measurement of finger posture and three-axis fingertip touch force using fingernail sensors. IEEE Trans. Robot. Autom. **20**(1), 26–35 (2004)

Fingertip Force Estimation Based on the Deformation of the Fingertip

Kibita Akihito, Toshio Tsuji and Yuichi Kurita

Abstract This paper describes a method of estimating the fingertip force and the application. We focus the deformation of fingertip contact surface when a fingertip is pressed and slightly slide on a rigid plate. The deformation changes depending on the exerted force on the contact surface. The proposed method allows the fingertip force estimation only from the contact images captured by a camera using FTIR method (Han 2005, [1]).

Keywords Force estimation · Fingertip contact · Eccentricity · FTIR · Haptic interface

1 Introduction

Humans can precisely control force at the fingertip. There are many receptors inside human fingertips, and they realize very high sensitivity of shear force applied on the fingertip. As a computer interface with good operationality, a force-based interface is a promising solution. However, it is difficult to realize the sensing capability of multiple axes force with a large sensing area by using a force sensor. In addition, multiple force sensors are required to discriminate force applied from different

K. Akihito (✉)
Graduate School of Engineering, Hiroshima University, 1-4-1 Kagamiyama, Higashihiroshima City, Hiroshima 739-8527, Japan
e-mail: kibita@bsys.hiroshima-u.ac.jp
URL: http://www.bsys.hiroshima-u.ac.jp/

T. Tsuji · Y. Kurita
Institute of Engineering, Hiroshima University, 1-4-1 Kagamiyama, Higashihiroshima City, Hiroshima 739-8527, Japan
e-mail: tsuji@bsys.hiroshima-u.ac.jp

Y. Kurita
e-mail: kurita@bsys.hiroshima-u.ac.jp

H. Kajimoto et al. (eds.), *Haptic Interaction*, Lecture Notes in Electrical Engineering 277, DOI 10.1007/978-4-431-55690-9_39

points. Some studies have been proposed to estimate the fingertip force without a force sensor. Marshall et al. [2] have developed an image-based force estimation method that can detect the fingertip force from the change in color of the nails. We have proposed a method to analytically calculate fingertip force based on the evaluation of the change in the area of the contact between the fingertip and a sensing surface [3]. In this paper, we describe a new fingertip force measurement device that can estimate the multiple axes force at multiple contact point.

2 Fingertip Force Calculation Method

2.1 Calculate Method of Vertical Force

The analytical relationship of the contact between an elastic object with radius of curvature of R_1 and R_1' and a rigid plate is modeled as the Hertzian contact [4]. The contact surface of the two objects has a shape of an ellipse with the major axis a and minor axis b. According to Hertz contact theory, the relationship between the vertical force and the contact area can be expressed by:

$$f_v = \frac{4\left(\frac{1}{R_1} + \frac{1}{R_1'}\right)}{3}\left(\frac{a}{\alpha}\right)^3 \frac{E_1 E_2}{E_1\left(1 - v_2^2\right) + E_2\left(1 - v_1^2\right)} \tag{1}$$

$$f_v = \frac{4\left(\frac{1}{R_1} + \frac{1}{R_1'}\right)}{3}\left(\frac{b}{\beta}\right)^3 \frac{E_1 E_2}{E_1\left(1 - v_2^2\right) + E_2\left(1 - v_1^2\right)} \tag{2}$$

where E is the Young's modulus, v is the Poisson's ratio, and α and β are the parameters determined by the angle of the plane containing the radius of curvature forms. The A and B are the parameters that show the initial gap between the two solid surface. This analytical relationship suggests that the vertical force f_v can be determined based on the measurable index a or b by camera images when the radius of the curvature and the parameter of the hardness of the object are known.

2.2 Calculate Method of Tangential Force

When shear force is applied on a flat plate from a fingertip, a minute displacement occurs on the fingertip contact surface. The tangential force f_t can be expressed by the following equation by a function of the eccentricity e_x, which is the index of the displacement [5], and the vertical force f_v:

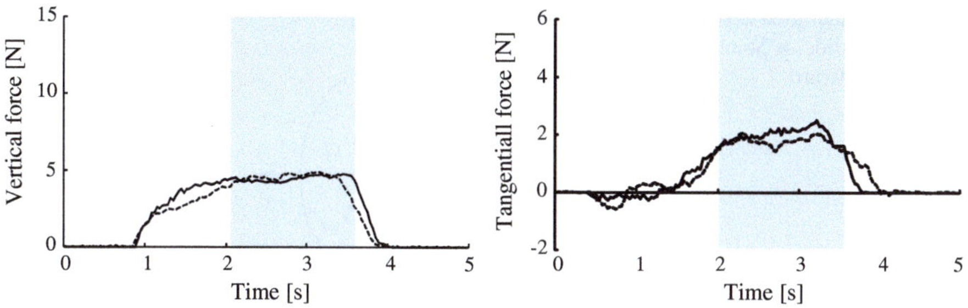

Fig. 1 Result of f_v (target force: f_v = 5.0 N, f_t = 2.0 N)

$$f_t = \mu f_v \left[1 - \left\{ 1 - \frac{8Ge_x a\pi(ab)^{\frac{2}{3}}}{3\mu f_v (2-\gamma)} \right\}^{\frac{3}{2}} \right] \tag{3}$$

where μ is the friction coefficient, γ is the dynamic coefficient of viscosity, R is the radius of the curvature of a fingertip and $G = E/2(1 + \gamma)$. When tangential force is applied on the fingertip, the contact area deforms as shown in Fig. 1. The eccentricity e_x along the x-axis is defined as follow:

$$e_x = \left(\frac{S_{t1} + S_{t4}}{S_t} - \frac{S_{t2} + S_{t3}}{S_t} \right) - \left(\frac{S_{s1} + S_{s4}}{S_s} - \frac{S_{s2} + S_{s3}}{S_s} \right) \tag{4}$$

where $S_{s1} - S_{s4}$ and $S_{t1} - S_{t4}$ are the divided areas shown in Fig. 1.

The force estimated by the proposed method and the true force measured by the force sensor are shown in Fig. 1. The mean error rates within the meshed area are also described in the figures. We confirmed that the proposed method can estimate fingertip force with the error rate of less than 15 % using above model equation.

3 Application

Figures 2 and 3 show the developed fingertip force measurement device that can estimate the multiple axes force at multiple contact point. The top of the device is a transparent acrylic plate, and the other outer covers are fabricated by a 3D printer. The contact area can be captured by a Web camera installed inside the device. The device is connected to a computer, and the image processing and the force estimation are done by the computer.

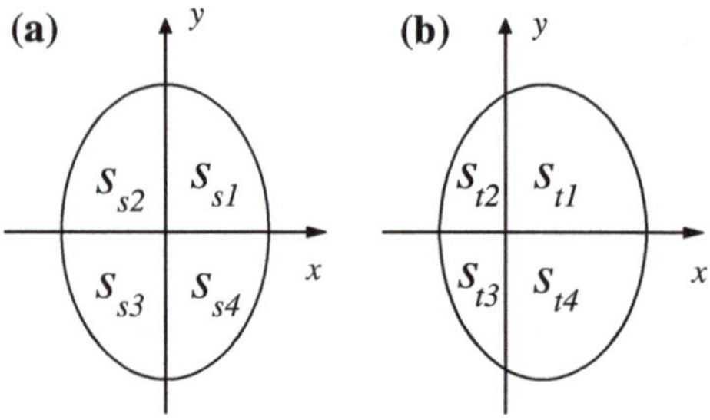

Fig. 2 Contact area before and after a slide. **a** State. **b** Slide rightward

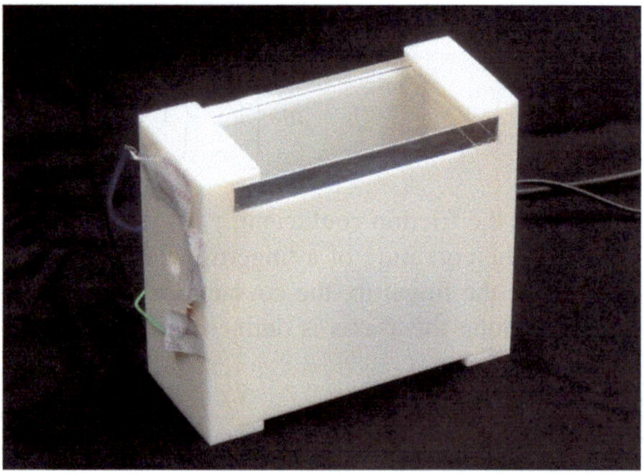

Fig. 3 Developed fingertip force measurement device

A user holds the device with two hands and puts his/her fingertips on the transparent plate (Fig. 4a). The contact area between the fingers and the plate is captured by a camera (Fig. 4b), and the contact radius is obtained by image processing. The obtained contact radius is used for the force estimation based on Eqs. (1), (2), and (3). The estimated force is shown on the computer screen (Fig. 4c).

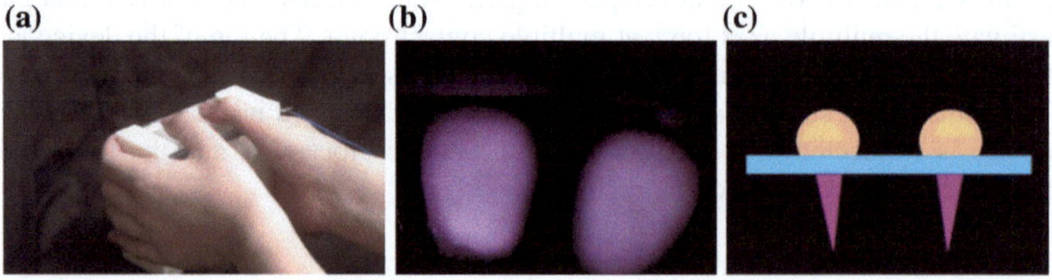

Fig. 4 Demonstration of the device. **a** Operate scene. **b** Captured image. **c** Estimated force direction

4 Conclusion

In this paper, we introduced the estimation method of the vertical and tangential force applied from the fingertip and proposed a fingertip force measurement device. As a future work, we plan to develop wider range type.

References

1. Han, J.Y.: Low-cost multi-touch sensing through frustrated total internal reflection. In: ACM UIST'05, pp. 115–118 (2005)
2. Marshall, J., Pridmore, T., Pound, M., Benford, S., Koleva, B.: Pressing the flesh: sensing multiple touch and finger pressure on arbitrary surfaces. In: Proceedings of Pervasive, pp. 38–55 (2008)
3. Kibita, A., Tsuji, T., Kurita, Y.: Fingertip force estimation based on the deformation of the fingertip. In: DHM2014 symposium program and paper abstracts, pp. 16, 20–22 May 2014
4. Tymoshenko, S.P.: Theory of Elasticity. McGraw-Hill Education, New York City (1970)
5. Kurita, Y., Ikeda, A., Ueda, J., Ogasawara, T.: A fingerprint pointing device utilizing the deformation of the fingertip during the incipient slip. IEEE Trans. Robot. **21**, 801–811 (2005)

Part V
Medical Application

Exoskeleton Simulator of Impaired Ankle: Simulation of Spasticity and Clonus

Hiroshi Okumura, Shogo Okamoto, Shun Ishikawa, Kaoru Isogai, Naomi Yanagihara-Yamada, Yasuhiro Akiyama and Yoji Yamada

Abstract We developed a prototype of an exoskeletal patient simulator that allows clinical trainees to experience and learn about ankle disorders related to hemiplegia. The exoskeleton exerts abnormal joint torques by tendon mechanisms while realizing complex ankle movements and realistic bone and skin features. Using this exoskeleton, we simulated the resistances of spasticity and clonus, which are typical symptoms of hemiplegia. We demonstrated these two types of simulated symptoms and showed their validity.

Keywords Patient simulator · Spasticity · Clonus · Physical therapy

1 Introduction

Physical therapists (PTs) manually examine a diseased joint to understand its clinical condition based on the dynamic joint resistance. Because unlicensed PT trainees have few opportunities to treat actual patients, some researchers have developed patient robots to simulate the symptoms of diseased joints for physical therapy training. For example, Grow et al. [3] and Park et al. [10] developed robotic simulators of spastic elbows. Kikuchi et al. [7, 8] simulated the spastic movements

H. Okumura (✉) · S. Okamoto · S. Ishikawa · Y. Akiyama · Y. Yamada
Department of Mechanical Engineering and Science, Nagoya University, Nagoya, Japan
e-mail: okumura.hiroshi@c.mbox.nagoya-u.ac.jp

K. Isogai
Department of Physical Therapy, Tokoha University, Hamamatsu, Japan

N. Yanagihara-Yamada
Department of Physical and Occupational Therapy, Nagoya University, Nagoya, Japan

H. Kajimoto et al. (eds.), *Haptic Interaction*, Lecture Notes
in Electrical Engineering 277, DOI 10.1007/978-4-431-55690-9_40

of the foot joint using a leg and foot robot. In the case of robotic patient simulators, the realism or human likelihood is typically limited because humans have multiple degrees of freedom (DOF) in their joints in addition to the skin and bone features. In order to address such issues, Ishikawa et al. [5, 6] proposed a framework for exoskeletal patient simulators. This concept is especially effective for complex human joints such as the feet [9].

The objective of this study is to develop a simulator for spasticity and clonus of the ankle for educational purposes. These two types of symptoms are significant and frequently developed for hemiplegia patients. The exoskeleton form and the tendon mechanism along a human muscle allow the trainees to experience high DOF ankle motion with a reduced number of actuators. PT trainees will be able to learn manual examination techniques to test for clonus and spasticity by using this simulator. Because our simulator has a tendon mechanism that applies torque only in the plantar direction, it seems difficult to simulate clonus, which involves both plantar and dorsal flexion movements of the ankle. However, utilizing manual force applied to the sole by a PT trainee, clonus can be simulated. We demonstrate the simulated clonus and spasticity function in this study.

2 Exoskeleton Mechanism to Simulate Spasticity and Ankle Clonus

Figure 1 shows a prototype of the exoskeletal simulator and a training scene. The exoskeleton was constituted mainly by a shoe and cuff fixed to the upper calf. A DC motor (RE35, Maxon motor, maximum continuous torque 97.2 mN m) with a 1/23 gear head was attached to the cuff. We used a servo amplifier (4-Q-DC ADS 50/5, Maxon motor) in the current control mode. A wire extended from the sole through an idler was wound by the DC motor, and its tip was attached to a pulley (ϕ = 20 mm). The wire was set along the tibialis posterior muscle to simulate the

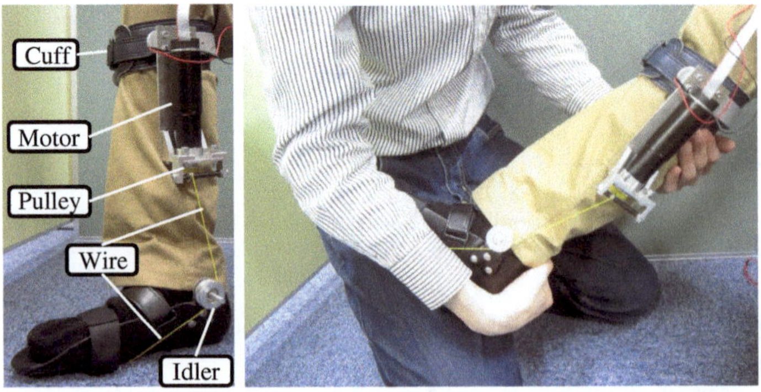

Fig. 1 Exoskeletal ankle simulator and training scene

disorders caused by the problems of this muscle. This wire-driven mechanism does not impede complex human ankle joint motions. One of the advantages of exoskeletal simulator is that there is no necessity to simulate the inherent mechanical impedance of human body. For example, the elasticity of the ankle muscle and tendon is presented by those of a wearer. The resistance of simulated symptom is superposed with that of the wearer's joint impedance.

3 Simulation of Spasticity

Spasticity is a disorder of the upper motor neuron, and it is characterized by a resistance proportional to the extension rate of the muscle [1]. The simulator presents the feeling of catch which PTs feel when they dorsiflex a foot of patients at higher rate than a certain value. The motor torque, τ_s, required to simulate spasticity was determined by

$$\tau_s = \begin{cases} \tau_0 & (\dot{\theta}(t) \leq \omega_0) \\ \tau_0 + c\dot{\theta}(t) & (\dot{\theta}(t) > \omega_0) \end{cases} \quad (1)$$

where c and $\dot{\theta}(t)$ are the coefficient of viscosity and angular velocity of the DC motor (the direction of ankle dorsiflex is the forward direction), respectively. If the angular velocity of the motor was less than or equal to ω_0, τ_s was constant at τ_0. Otherwise, τ_s was proportional to the angular velocity when $\dot{\theta}(t) > \omega_0$. This angular velocity, ω_0, represents the stretch reflex threshold [2]. We set each parameter as follows: $\tau_0 = 0.072$ N m, $c = 9 \times 10^{-3}$ N m s/rad, and $\omega_0 = 4.0$ rad/s.

We measured the resistance forces of the simulated spasticity in the setup shown in Fig. 2. A force sensor (USL06-H5-200N-C, Tec Gihan Co. Ltd, Japan.) was installed on the fore part of the sole such that it covered the load path over the fore foot. A PT dorsiflexed the ankle of the exoskeleton wearer, in a way similar to how it is usually performed in a clinical setting. Figure 3 shows a sample of the measured force and $\theta(t)$. The ankle was dorsiflexed quickly at $t = 3$, and an abrupt rise in the force was observed during this phase. Such a sharp increase in the interaction force was regarded as a typical velocity-dependent resistance of a spastic ankle.

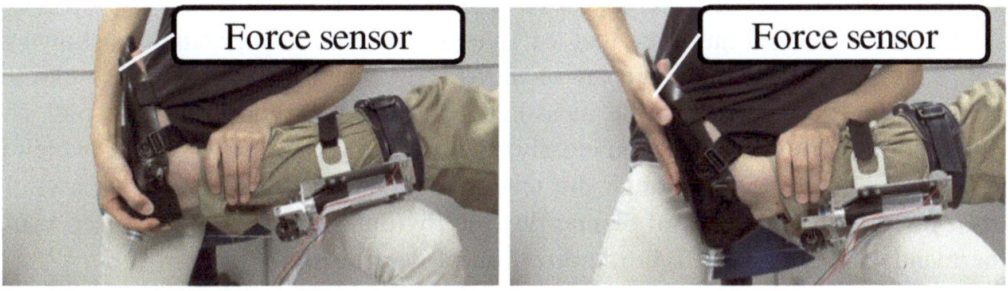

Fig. 2 Measurement setup; *left* spasticity test and *right* clonus test

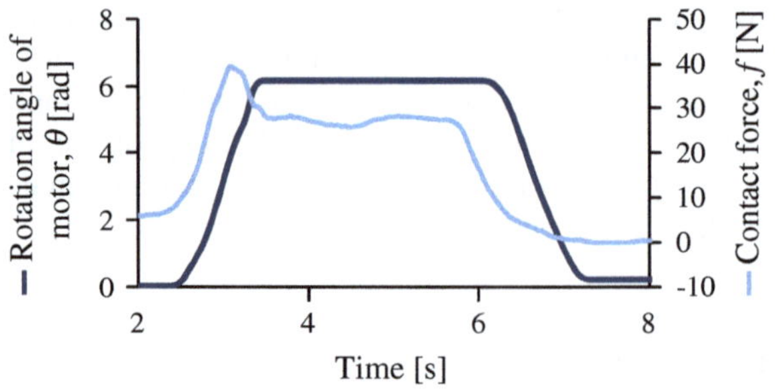

Fig. 3 Resistance force of the simulated spasticity. Force against the dorsi exion. Angle of the DC motor with $\theta = 0$ at full plantar exion

4 Simulation of Clonus

Clonus is also caused by disorders of the upper motor neuron, and it is characterized by an involuntary rhythmic contraction of muscles [4]. To test for ankle clonus, PTs quickly flex the ankle in the dorsal direction. When clonus occurs, PTs receive rhythmical and repetitive plantar and dorsal flexion of the ankle. The motor torque, τ_c, to simulate ankle clonus was determined by

$$\tau_c = \begin{cases} \tau_1 & (\dot{\theta}(t) \leq \omega_t) \\ \tau_2 & (\dot{\theta}(t) > \omega_t, \quad nt_s \leq t < (n+1)t_s, \; n = 0, 2, 4, \ldots 20) \\ 0 & (\dot{\theta}(t) > \omega_t, \quad nt_s \leq t < (n+1)t_s, \; n = 1, 3, 5, \ldots 21) \end{cases} \tag{2}$$

where $\tau_{1,2}$ and t_s are the constant resistance torques and rhythmic period of the ankle movements, respectively. When $\dot{\theta}(t)$ was smaller than ω_t, the reflective contraction did not occur and a weak constant resistance of τ_1 was set. When the ankle was dorsiflexed quickly and $\dot{\theta}(t)$ became larger than ω_t, a large resistance of τ_2 was set. τ_2 was sufficiently large to plantarflex the ankle joint, opposing the PT's manual force. This strong reflexive muscle contracture was then turned on and off at a period of $2t_s$. As a result, rhythmic plantar and dorsal flexion movements were produced. To express the moderate disappearance of clonus, we set each parameter as follows: $\tau_1 = 0.072$ N m, $t_s = 0.15$ s, $\tau_2 = 0.28$ N m for $n = 0, 2,\ldots 8$, 0.18 N m for $n = 10$, 0.14 N m for $n = 12$, and 0.12 N m for $n = 14$.

We also measured the force that a PTs experience on his/her hand, as shown in Fig. 2. A PT pushed the fore part of the sole manually to test for ankle clonus. Figure 4 shows a sample of the measured contact forces normal to the sole. The rhythmic motions were observed clearly at approximately 2–4 s at 3.4 Hz, which is in a typical range [4]. Although the tendon mechanism of the exoskeleton could exert force in the direction of plantar flexion, rhythmic clonus was simulated using the manual force of the PT in the direction of dorsiflexion.

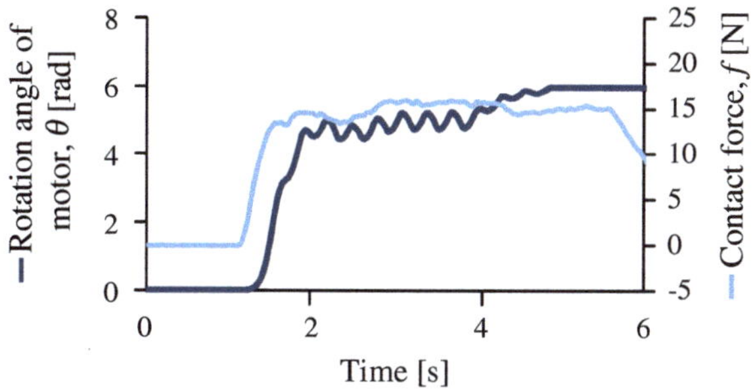

Fig. 4 Measurement results for ankle clonus (ankle was quickly dorsiflexed)

5 Conclusion

In this study, we expanded the functions of the exoskeletal ankle simulator to simulate spasticity and clonus. Because our exoskeleton was based on a tendon mechanism, it was challenging to simulate clonus, which involves both plantar and dorsal flexion movements. However, utilizing the manual force applied to the sole by a PT trainee, we could realize the typical abnormality of clonus. During manual examination toward each simulated symptom, resistance pattern characteristics of clonus and spasticity were observed between the wearer's fore foot and the trainee's hand or arm, suggesting the authenticity of the simulated symptoms. PT trainees will be able to learn manual examination techniques to test for clonus and spasticity by using this simulator.

Acknowledgment This work was in part supported by Naito Foundation.

References

1. Dietz, V., Sinkjaer, T.: Spastic movement disorder: impaired reflex function and altered muscle mechanics. Lancet Neurol. **6**, 725–733 (2007)
2. Galota, A., Feldman, A.G., Levin, M.F.: Spasticity measurement based on tonic stretch reflex threshold in stroke using a portable device. Clin. Neurophysiol. **119**(10), 2329–2337 (2008)
3. Grow, D.I., Wu, M., Locastro, M.J., Arora, S.K., Bastian, A.J., Okamura, A.M.: Haptic simulation of elbow joint spasticity. In: Proceedings of IEEE Symposium on Haptic Interfaces for Virtual Environments and Teleoperator Systems, pp. 475–476 (2008)
4. Hidler, J.M., Rymer, W.Z.: A simulation study of reflex instability in spasticity: origins of clonus. IEEE Trans. Rehabil. Eng. **7**(3), 327–340 (1999)
5. Ishikawa, S., Okamoto, S., Akiyama, Y., Isogai, K., Yamada, Y., Hara, S.: Wearable dummy to simulate joint impairment: model for the discontinuous friction resistance due to arthritis. In: Proceedings of IEEE International Conference on Robotics and Biomimetics, pp. 1409–1414 (2012)
6. Ishikawa, S., Okamoto, S., Isogai, K., Akiyama, Y., Yanagihara, N., Yamada, Y.: Wearable dummy to simulate joint impairment: severity-based assessment of simulated spasticity of knee joint. In: Proceedings of the IEEE/SICE International Symposium on System Integration, pp. 300–305 (2013)

7. Kikuchi, T., Oda, K., Furusho, J.: Leg-robot for demonstration of spastic movements of brain-injured patients with compact magnetorheological uid clutch. Adv. Rob. **24**, 671–686 (2010)
8. Kikuchi, T., Oda, K., Yamaguchi, S., Furusho, J.: Leg-robot with MR clutch to realize virtual spastic movements. J. Intell. Mater. Syst. Struct. **21**, 1523–1529 (2010)
9. Okumura, H., Okamoto, S., Ishikawa, S., Akiyama, Y., Isogai, K., Hirano, Y., Yamada, Y.: Wearable dummy to simulate equinovarus for training of physical therapists. In: Proceedings of SICE Annual Conference, pp. 2272–2277 (2013)
10. Park, H.S., Kim, J., Damiano, D.L.: Development of a haptic elbow spasticity simulator (HESS) for improving accuracy and reliability of clinical assessment of spasticity. IEEE Trans. Neural Syst. Rehabil. Eng. **20**, 361–370 (2012)

A Surgery Simulator Using an Optimized Space and Time Adaptive Deformation Simulation on GPU

Ryo Kuriki, Kazuyoshi Tagawa and Hiromi T. Tanaka

Abstract In this study, we propose an interactive surgery simulator using an optimized space and time adaptive (online re-mesh and multi-rate) deformation simulation on graphics processing unit (GPU). In our previous approach, relative large processing time was required for memory transfer between CPU and GPU, because updating of a node list was not performed after the re-meshing. We solved this problem by updating the node list with a fast and minimum updating procedure.

Keywords Elastic object · Online re-mesh · Multi-rate simulation · GPU · Surgery training system

1 Introduction

A computationally efficient deformation simulation approach for elastic objects is required of surgical simulators, movies, games and so on. In particular, in cases of presenting stable reaction force to users when interacting with the elastic objects, a higher update rate (about several hundred Hz) of the deformation simulation and the calculation of reaction force is required.

Of course, several acceleration approaches have been proposed. One potential approach is a parallel computation on graphics processing unit (GPU). Rasmusson

R. Kuriki (✉) · K. Tagawa · H.T. Tanaka
Ritsumeikan University, Shiga, Japan
e-mail: kuriki@cv.ci.ritsumei.ac.jp

K. Tagawa
e-mail: tagawa@cv.ci.ritsumei.ac.jp

H.T. Tanaka
e-mail: hiromi@cv.ci.ritsumei.ac.jp

H. Kajimoto et al. (eds.), *Haptic Interaction*, Lecture Notes
in Electrical Engineering 277, DOI 10.1007/978-4-431-55690-9_41

et al. [1] proposed an efficient parallel computation approach for simulating a mass-spring model on GPU. In their approach, a connection table and a node list are used for the simulation. To realize further acceleration, we adapted this approach to a space and time adaptive (online re-mesh and multi-rate) simulation [2]. In the adaptive simulation, due to the numbers of nodes and springs may be changed after re-meshing, updating of the connection table and the node list is necessary. Furthermore, especially in GPU, optimizing memory access patterns is very important for efficient computations. Therefore, we proposed a fast and minimum updating (defragmenting and sorting) approach for the connection table. However, large processing time for memory transfer between CPU and GPU was required, because updating of the node list was not performed.

In this study, to solve above problem, a fast and minimum updating approach for the node list is proposed. We implement this approach into our laparoscopic surgery simulator and show effectiveness of proposed approach.

2 Optimized Space and Time Adaptive Deformation Simulation on GPU

In our previous paper, a fast and minimum updating approach for the connection table was proposed [2]. However, in this approach, node list was in fragments as shown in Fig. 1. Therefore, large processing time for memory transfer (cudaMemcpy) of entire list between CPU and GPU was required.

In this research, we propose an updating approach for the node list. By using this approach, the node list is also defragmented and sorted by time step as shown in Fig. 2.

Figure 3 shows an example of an updating procedure for the node list in new node addition, where Δt *is* a time step of each node for the multi-rate deformation simulation, (X, Y, Z) is position of each node. After this updating, modification of the connection table is required because indices of copied nodes are changed and

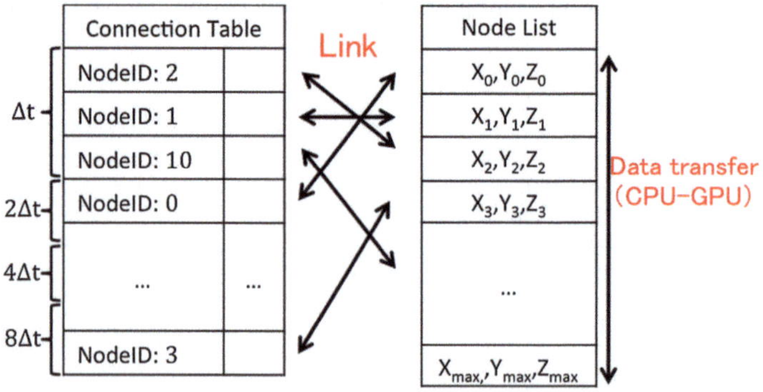

Fig. 1 Connection table and node list in previous approach

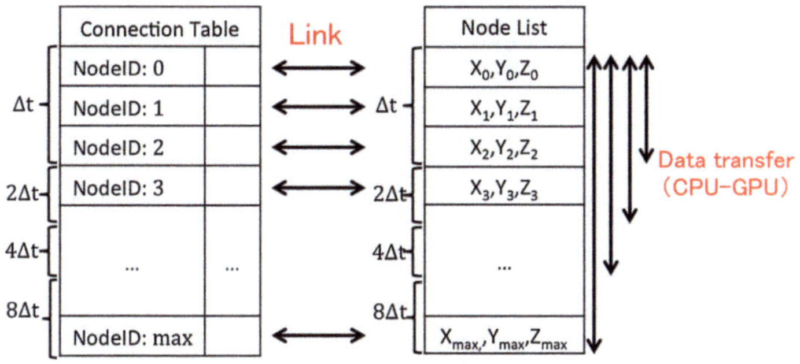

Fig. 2 Connection table and node list in proposed approach

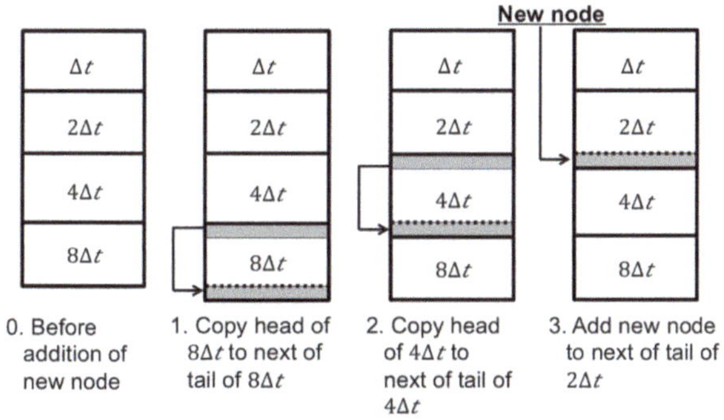

Fig. 3 Example of updating of node list in new node addition

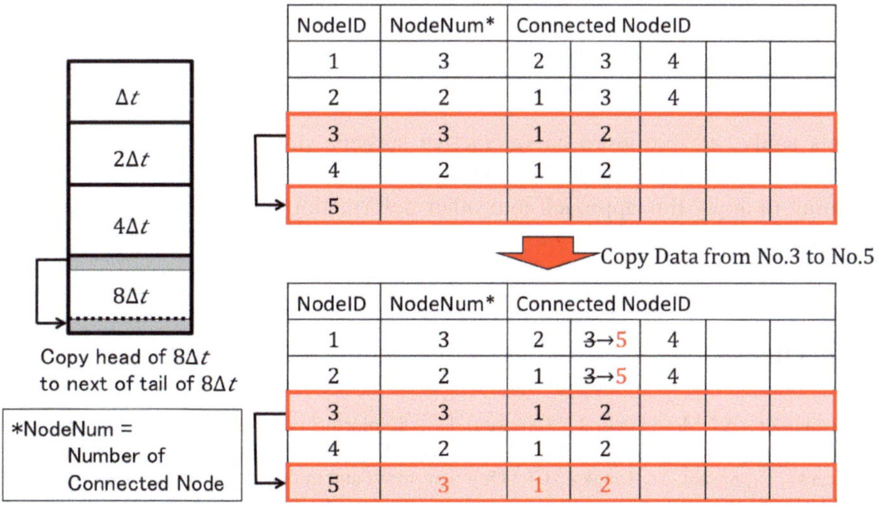

Fig. 4 Example of modification of connection list. In this example, index of a node was changed 3–5

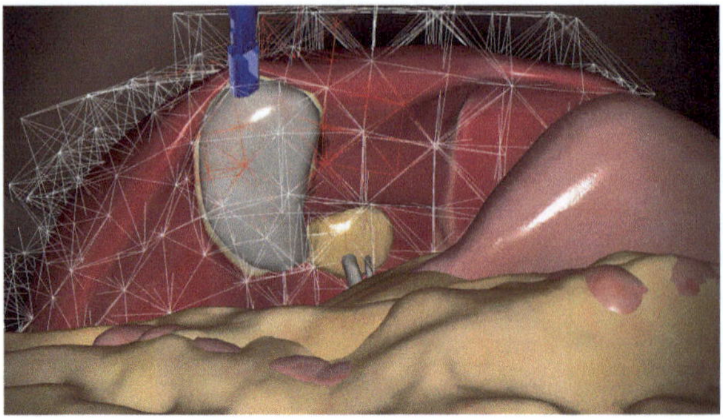

Fig. 5 Simulation result

indices of connected nodes are stored in the connection table. Figure 4 shows an example of this modification. By using this approach, the node list is defragmented and sorted by their time steps. As a result, efficient deformation simulation on GPU and fast memory transfer between CPU and GPU become possible.

3 Results

We implemented the proposed approach into our developing surgery simulator. Figure 5 shows a simulation result. The user grasped and pulled a gallbladder. In current implementation, processing time of deformation simulation and memory transfer per one step was about 0.2 ms.

4 Conclusions

In this study, we proposed an interactive surgery simulator using an optimized space and time adaptive deformation simulation on GPU. In the future, we are planning to apply this approach into other deformation model using finite element method.

References

1. Rasmusson, A., Mosegaard, J., Sørensen, T.S.: Exploring parallel algorithms for volumetric mass-spring-damper models in CUDA. In: Proceedings of ISBMS, pp. 49–58 (2008)
2. Tagawa, K., Sasaki, Y., Tanaka, H.T.: Online re-mesh and multi-rate deformation simulation by GPU for haptic interaction with large scale elastic objects. In: Proceedings of IEEE Haptics, pp. 531–538 (2012)

Hierarchical Examination of Colliding Points Between Rigid and Deformable Objects

Mary-Clare Dy, Kazuyoshi Tagawa, Hiromi T. Tanaka and Masaru Komori

Abstract Virtual reality-based simulators can be used as a tool for surgical training, provided that such simulators can produce realistic tool-to-organ interactions. To reproduce such interactions, we are developing a laparoscopic simulator with haptic feedback. However, the current simulator is designed to have one point at each tool tip to make contact with the organs. We propose a method to improve the simulator to detect the collision not only on the tool tips, but on the tool rods as well, using a hierarchical approach.

Keywords Laparoscopic simulator · Haptic feedback · Collision detection · Hierarchical binary representation · Deformable objects

1 Introduction

Minimally invasive surgery (MIS) or laparoscopic surgery emerged as an alternative procedure for some open surgeries. Better pain management and faster recovery time are just some of the advantages experienced by patients. However, this type of surgery requires extensive technical training, because it requires mastering different motor skills, and hand–eye coordination. Surgeons experience minimised sense of touch as well. To support such surgical training, a laparoscopic simulator with haptic feedback is being developed. In this paper, we focus on the interaction between rigid objects (tools) and deformable objects (liver, gallbladder, fat). The simulator uses a

M.-C. Dy (✉) · K. Tagawa · H.T. Tanaka
Ritsumeikan University, Kyoto, Japan
e-mail: claredy@belle.shiga-med.ac.jp

M.-C. Dy · M. Komori
Shiga University of Medical Science, Ōtsu, Japan

H. Kajimoto et al. (eds.), *Haptic Interaction*, Lecture Notes
in Electrical Engineering 277, DOI 10.1007/978-4-431-55690-9_42

point on each tool tip to detect collisions with the organs. But in real situations, surgeons also use the tool rods to manipulate the organs. We are now investigating how to extend the simulator to include multiple collision detections and its consequent deformation on the organs.

2 Current Laparoscopic Simulator: Background and Problem

We use a mass-spring-damper (MSD) model and two types of nonlinear finite element method (FEM) model for deformation simulation. An online remesh deformation simulation and a parallel computing by GPU accelerate the deformation simulation [1]. A ray-to-triangle intersection algorithm detects collisions [2] between the virtual tools and organs. When the tool tip probes the organ in a downward motion, the collision detection algorithm calculates the collision point and depth of collision. The rendering scheme then corrects the interpenetration of the tool into the organ. Finally, the haptic devices provide feedback to signal the user that collision has occurred.

The simulator can realistically show the interaction between the tool tips and organs. However, interpenetration between the tool rods and organs occurs as shown in Fig. 1. We first performed preliminary tests to see whether the simulator can handle more collision points. Although the simulator was able to detect collisions when multiple points were placed along the tool rods, the sequential examination of possible collision increased the cost of computation and caused the organ deformation to slow down. Not to mention, haptic jittering occurred when there was a sudden change in the collision point position. We surmised that faster collision detection algorithm must be developed that can match the update rate of the haptic device to provide smooth and realistic feedback. In this paper, we will focus on the collision detection of multiple points.

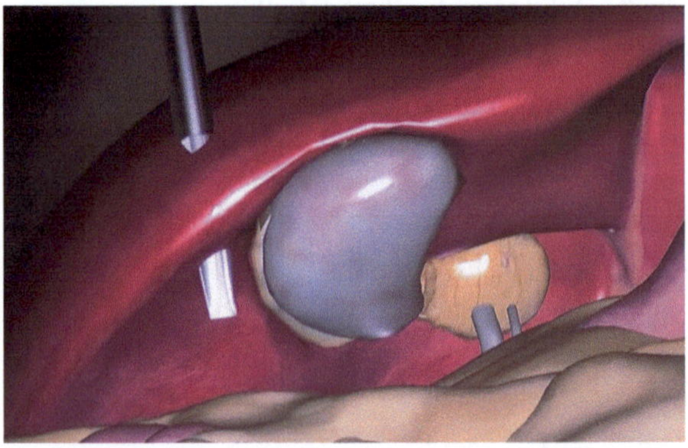

Fig. 1 Interpenetration of tool into the organ

3 Proposed Modification of Collision Detection and Deformation Modelling

3.1 Collision Scheme for Interacting Objects

A number of *reduced* deformable objects can be simulated using [3] an approach, wherein one object is modelled using a floating-point signed-distance field and the other as pointshells organised into a nested multi-resolution hierarchy. The latter means that each point primitive will have different sizes of spherical bounding volumes with respect to the point distribution. The points and their corresponding pre-computed normals are sampled from a near uniform distribution. A line-based approach was proposed by Maciel and De [4], wherein the tool rod is composed of a line segment with a dynamic point. The dynamic point can change its position to determine the closest point to the colliding deformable object as long as it lies within the line segment. The line segment could be decomposed into smaller segments each with their own dynamic points to address nonconvex objects. When the closest distance between the line segment and the other object is found, a triangle strip is created based on the length of the line segment that contains the dynamic point. This triangle strip is used to deform the other object. The pointshell scheme can be adapted to our simulator; however, we must optimise the scheme such that fewer points will be traversed. Using dynamic point, method could decrease the number of points; however, the endpoints of the line segment cannot be altered to provide a more accurate area of deformation on the other object. Also, dynamic point was not yet successfully implemented in a laparoscopic simulator. Based on these works, we designed our method for our simulator. An image representing our method is shown in Fig. 2.

In our study, the rod will be enclosed with an oriented bounding box (OBB), while the organs are enclosed in an adaptive grid [5]. The bounding volumes of

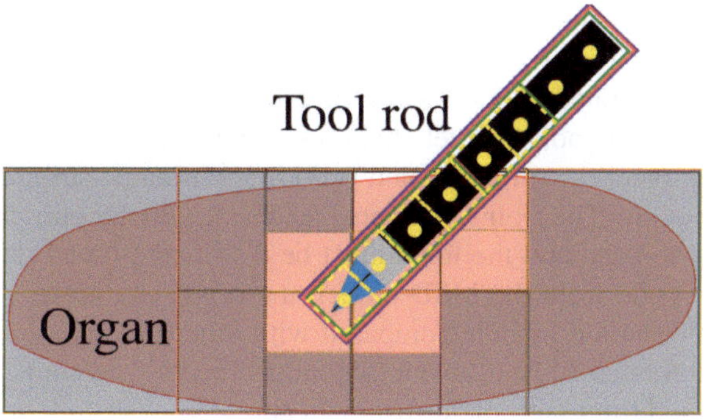

Fig. 2 Laparoscopic tool and organ representation. Oriented bounding boxes on the tool and adaptive grid on the organ are traversed hierarchically to detect possible collision between the two objects

both objects will be checked for possible collisions and then bisected hierarchically until the precise area, or point, of collision is detected. Hierarchical bounding volume technique using OBB can effectively decrease the computation costs as opposed to other bounding volumes [6]. Next, we define the object primitives: we can specify the minimum length of the leaf box, and each box will contain one contact point positioned at the centroid. This arrangement lessens the number of points that make up the rod, while ensuring that the whole rod is taken into account. The organs are made up of triangle patches. In the pre-computation, the adaptive grid stores the information of neighbouring tetrahedrons. We will also include a list to store the neighbouring triangle patches of the organ surfaces. Third step, when the leaf nodes of the organ's grid are reached, we compute the maximum length of the rod that will be in contact with the organ. As mentioned, the endpoints of the contacting part will be determined hierarchically. Then, from each of the endpoints, a ray will be casted towards the surface of the organ. The intersected triangle patches will be used as the endpoints for creating a triangle strip, similar to [4], but with a modification: we determine the closest point between the region of interest, that is the areas enclosed within the rod endpoints, and the triangle strip. This is done by projecting rays, first from the midpoint of the rod's endpoints, towards each triangle patch of the created triangle strip. The measured distances between the midpoint and each triangle patch are compared, and the rod point and triangle patch with the shortest distance are saved. The rod area is recursively bisected until the position with the shortest distance between the rod and organ is determined. The resulting point of the rod and triangle patch of the organ serves as the main haptic feedback point, and origin of deformation, respectively.

3.2 Organ Deformation and Haptic Feedback

The deformation based on multiple external forces applied by the rod caused a computational bottleneck in Maciel and De [4]. In our simulator, although it also uses MSD model, it can already handle multiple applied external forces. But to ascertain that the deformation will coincide with the haptic feedback, the triangle strip must be incorporated in our system. In this case, the adaptive grid can provide the information about adjacent tetrahedrons. And then, an additional table for adjacent triangle patches will be created to form the triangle strip, as mentioned in the previous section. The point of haptic feedback will be chosen based on the closest point of contact, and the force must be distributed based on the number of contact points. Aside from the MSD model, the organ can be represented by corotated-FEM deformation model. Compared with other deformation models within the simulator, corotated-FEM is able to accurately render the geometric nonlinearity that often occurs during large deformations [7].

References

1. Tagawa, K., Sasaki, Y., Tanaka, H.T.: Online re-mesh and multi-rate deformation simulation by GPU for haptic interaction with large scale elastic objects. In: Proceedings of IEEE Haptics, pp. 533–540 (2012)
2. Möller, T., Trumbone, B.: Fast, minimum storage ray-triangle intersection. In: ACM SIGGRAPH 2005 Courses, Proceedings of SIGGRAPH, ACM, Article 7 (2005)
3. Barbič, J., James, D.L.: Time-critical distributed contact for 6-DoF haptic rendering of adaptively sampled reduced deformable models. In: Proceedings of Eurographics/ACM SIGGRAPH Symposium on Computer Animation, Eurographics Association, pp. 171–180 (2007)
4. Maciel, A., De, S.: An efficient dynamic point algorithm for line-based collision detection in real time virtual environment involving haptics. Comp. Anim. Virtual Worlds **19**(2), 151–163 (2008)
5. Tanaka, H.T., Takama, Y., Wakabayashi, H.: Accuracy-based sampling and reconstruction with adaptive grid for parallel hierarchical tetrahedrization. In: Proceedings of 2003 Eurographics/IEEE TVCG Workshop on Volume Graphics, ACM, pp. 79–86 (2003)
6. Gottschalk, S., Lin, M.C., Manocha, D.: OBBTree: a hierarchical structure for rapid interference detection. In: Proceedings of the 23rd Annual Conference on Computer Graphics and Interactive Techniques, ACM/SIGGRAPH, pp. 171–180 (1996)
7. Tagawa, K., Yamada, T., Tanaka, H.T.: A rectangular tetrahedral adaptive mesh based corotated finite element model for interactive soft tissue simulation. In: Proceedings of Engineering in Medicine and Biology Society (EMBC), 2013 35th Annual International Conference of the IEEE, IEEE, pp. 7164–7167 (2013)

Wearable Robot for Simulating Knee Disorders in the Training of Manual Examination Techniques

Shun Ishikawa, Shogo Okamoto, Kaoru Isogai, Naomi Yanagihara-Yamada, Yasuhiro Akiyama, Yujiro Kawasaki and Yoji Yamada

Abstract This study addresses a haptic simulator of diseased knee joints that is intended for the use in the training of manual examination techniques used in physical therapy. These techniques involve clinicians moving the impaired joint of a patient to test the dynamic joint resistance caused by a condition. Our exoskeleton robot simulates five types of common knee problems. The simulated symptoms were checked in terms of subjective similarity to actual symptoms that physical therapists experience in their work. Three of the five symptoms could be correctly identified, whereas the other two were found to require further tuning. The proposed patient simulator could improve the training of manual examination techniques by being able to simulate a variety of symptoms.

Keywords Patient simulator · Physical therapy · Manual examination

1 Introduction

Physical therapists manually examine the biomechanical and neural responses of the limbs of their patients to understand the nature of their disorders. In these examinations, the therapists test the dynamic joint resistance, as well as the range of joint

---

Shun Ishikawa: This study was in part supported by Naito Foundation.

S. Ishikawa · S. Okamoto (✉) · K. Isogai · Y. Akiyama · Y. Kawasaki · Y. Yamada
Department of Mechanical Science and Engineering, Nagoya University, Nagoya, Japan
e-mail: okamoto-shogo@mech.nagoya-u.ac.jp

K. Isogai
Department of Physical Therapy, Tokoha University, Hamamatsu, Japan

N. Yanagihara-Yamada
Department of Physical and Occupational Therapy, Nagoya University, Nagoya, Japan

H. Kajimoto et al. (eds.), *Haptic Interaction*, Lecture Notes in Electrical Engineering 277, DOI 10.1007/978-4-431-55690-9_43

motion, when the impaired limbs are moved passively. Although a therapist requires clinical experience to master these manual examination techniques, educational facilities rarely have the means of giving trainee therapists experience with real patients. To solve this issue, we have been developing patient simulators, which will allow trainee therapists to experience the haptic characteristics of diseased joints [2, 3]. A unique aspect of our approach is that we use an exoskeleton robot to simulate joint impairment, rather than taking the robotic patient approach of earlier schemes [4, 6]. In our approach, a healthy person wears the exoskeleton, which simulates the abnormal behavior of a patient's limb. Since the simulator incorporates the part of an actual human body, a high level of realism can be achieved, including the complexity of human joint motion, as well as the bone and skin features of the human body.

The objective of this study is to demonstrate and validate the authenticity of a variety of simulated knee symptoms using a knee exoskeleton robot, building on our previous work, whereas we simulated a spastic knee joint and crepitus [2, 3]. In this study, we also set out to simulate the symptoms of contracture, lead-pipe rigidity, cogwheel rigidity, velocity-dependent resistance, and bony ankylosis, which are common symptoms that physical therapists are likely to treat in a clinical setting, and which are described in the following section. To test the authenticity of the simulated symptoms, we invited practicing physical therapists to participate in a symptom-classification experiment.

2 Wearable Knee Impairment Simulator

Figure 1 shows a mock patient, wearing the knee exoskeleton, and a trainee therapist whose task is to specify the knee joint impairment. The wearer relaxes and does not invoke any voluntary motions, while the trainee flexes or stretches the wearer's leg to test the dynamic force feedback. This is undertaken in an educational facility, where the trainees take turns in playing the role of patient when learning these manual techniques.

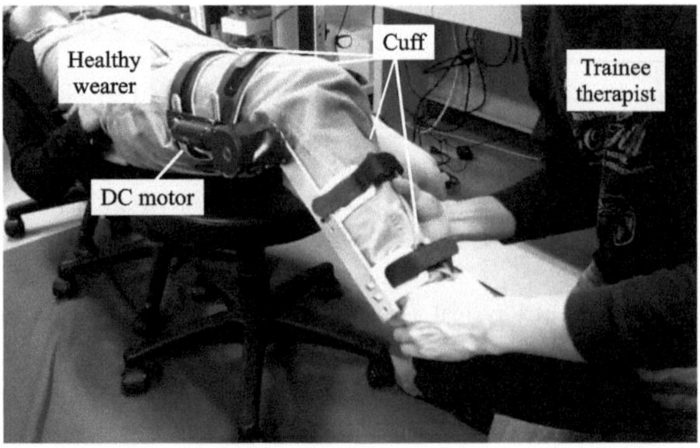

Fig. 1 Exoskeleton-type knee patient simulator. The healthy wearer (*on the left*) plays the role of patient. The trainee (*on the right*) practices the examination techniques

The knee robot consists of two links, fastened to the wearer's femoral and lower thighs, and one DC motor (RE35, Maxon, stalling torque = 949 mN m, 1/86 gearing) which exerts a torque around the knee joint through a pair of bevel gears. The DC motor is controlled by a microcomputer operating at 5 kHz through a voltage–current converter (4-Q-DC ADS 50/5, Maxon). This system, which was developed as a part of another study, was able to measure the joint angle and one-dimensional force applied to the tibia link.

3 Simulated Symptoms

We simulated five different symptoms. Three of these, namely contracture, lead-pipe rigidity, and velocity-dependent resistance, were also simulated in some earlier studies using robotic patient simulators. We focused on the authenticity and feasibility of the five types of simulated symptoms produced by the wearable joint simulator, rather than on developing sophisticated algorithms. Here, we briefly describe each symptom and how they were simulated. Each symptom involves gain parameters, which determine the severity of a symptom or its characteristics. In a latter experiment, these parameters were moderately set by the coauthors of this article, both of whom are physical therapists, such that each symptom was barely discernible.

Lead-pipe rigidity Lead-pipe rigidity, which typically accompanies Parkinson's disease, is characterized by a constant stiffness of the joint. This stiffened joint can be simulated by providing a constant resistance against passive joint motion. We set the joint resistance torque of the lead-pipe rigidity to 0.5 N m.

Cogwheel rigidity Cogwheel rigidity is similar to lead-pipe rigidity; however, it is characterized by a nonuniform resistance. We modeled this symptom by

$$\tau_{\mathrm{cog}} = \begin{cases} 1.0\ \mathrm{N\ m} & \\ 0.5\ \mathrm{N\ m} & \left(\frac{\pi}{9} < \theta < \frac{5\pi}{36}, \frac{\pi}{6} < \theta < \frac{7\pi}{36}, \frac{2\pi}{9} < \theta < \frac{\pi}{4}\right) \end{cases} \quad (1)$$

where the fully extended angle was 0 rad. As a result, intermittent joint stiffness is experienced by the trainee as he/she manipulates the knee.

Contracture Contraction involves chronically stiffened tendons or muscles constraining the joint movement. This limitation of the range of motion is typically modeled by using a spring coefficient. We also simulated contracture as an elastic resistance (5.7 N m/rad) that acted only during knee flexion. In addition, to provide a resistance-free range, the elastic resistance was not exerted until the joint angle reached $\pi/18$ rad.

Bony ankylosis An abnormal bony deformation or an artificial joint also limits the range of motion of the joint. The therapist can sense the contact between the nonelastic bones at the end of the range of motion. We simulated the resistance of this symptom by using a large spring constant of 28.5 N m/rad. The free motion range was set to 0 to $2\pi/3$ rad.

Velocity-dependent resistance A velocity-dependent joint resistance is caused by an abnormal stretch reflex of a joint and is also referred to as spasticity. This type of resistance was simulated as a viscosity relative to an angular velocity in some earlier studies [1, 5]. We also followed their approach and determined the resistance torque from

$$\tau_{\mathrm{vel}} = \begin{cases} c\dot{\theta}(t) & (\dot{\theta}(t) \geq 0) \\ 0 & (\text{otherwise}) \end{cases} \tag{2}$$

where the viscosity coefficient was set to 1.5 Nm s/rad.

4 Classification Experiment to Check the Authenticity of the Simulated Symptoms and the Knee Patient Simulator

4.1 Methods and Tasks

We conducted a classification experiment in which physical therapists experienced and classified the simulated symptoms so as to validate the authenticity of our simulator. All of the procedures were approved by the IRB of the Engineering School of Nagoya University.

A healthy person wore the exoskeleton knee robot on the right leg and lay on a bed in a relaxed posture as shown in Fig. 1. Nine physical therapists, each with more than 5 years of clinical experience, took part in the experiment after providing their written informed consent. None of them had any experience of robotic joint impairment simulators. Before the main task, each was shown a list of the symptoms to be simulated and then actually experienced all of them. At this point, they were not provided with any feedback on the presented symptoms. In the main task, each participant was allowed to freely test each symptom for a maximum of one minute, before providing the name of the symptom that he/she believed was being simulated. The experiment thus involved a five-alternative forced-choice task. The simulated symptoms were presented in random order with each being experienced five times in total. Hence, each participant performed twenty-five trials.

4.2 Results—Positive Answer Ratios

Figure 2 shows the averages and standard deviations of the answer ratios at which the presented and classified symptoms matched. Lead-pipe rigidity, cogwheel rigidity,

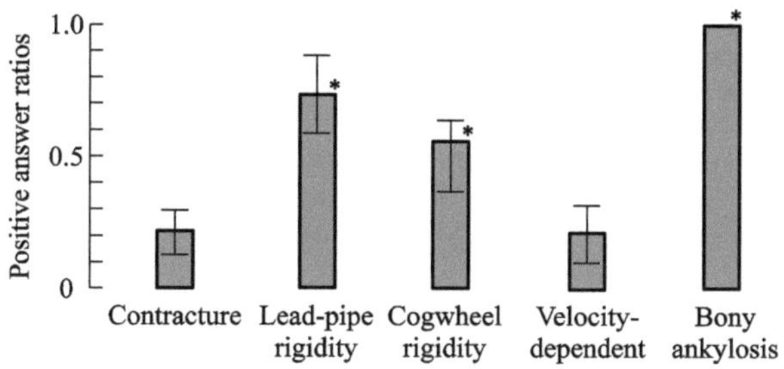

Fig. 2 Average positive classification ratios of simulated symptoms and standard deviations among the participants. Here, * indicates a statistically significant difference from the chance (0.2) with $p < 0.05$ of two-tailed t test

and bony ankylosis were classified as expected with statistically significant probabilities. In particular, ankylosis was always classified correctly. However, the positive answer ratios for the simulated contracture and velocity-dependent resistance were almost the chance. Velocity-dependent resistance was often confused with lead-pipe rigidity (0.31 on average) and contracture (0.33 on average). According to the introspective reports provided by the participants after the main task, the velocity-dependent resistance was too small to differentiate from the other two symptoms. Moreover, contracture was often wrongly classified as velocity-dependent resistance (0.40 on average). The participants could not differentiate between these two types of symptoms possibly because the force feedback gains were set to low values.

5 Conclusion

In this study, we simulated several types of common knee disorders using an exoskeleton robot for providing training in manual examination techniques used by physical therapists. Because the wearable simulator for joint impairment is well balanced in terms of realism and cost, its practical use would allow the simulation of a greater number of symptoms. The five symptoms, namely contracture, lead-pipe rigidity, cogwheel rigidity, bony ankylosis, and velocity-dependent resistance, were simulated and evaluated. Although the gain that determines the prominence of the joint resistance has bet to be tuned to allow the device's use in an educational facility, however, the simulated symptoms could be correctly classified in a test involving nine physical therapists.

References

1. Grow, D.I., Wu, M., Locastro, M.J., Arora, S.K., Bastian, A.J., Okamura, A.M.: Haptic simulation of elbow joint spasticity. In: Proceedings of the IEEE Symposium on Haptic Interfaces for Virtual Environments and Teleoperator Systems, pp. 475–476 (2008)

2. Ishikawa, S., Okamoto, S., Akiyama, Y., Isogai, K., Yamada, Y., Hara, S.: Wearable dummy to simulate joint impairment: model for the discontinuous friction resistance due to arthritis. In: Proceedings of the IEEE International Conference on Robotics and Biomimetics, pp. 1409–1414 (2012)
3. Ishikawa, S., Okamoto, S., Isogai, K., Akiyama, Y., Yanagihara, N., Yamada, Y.: Wearable dummy to simulate joint impairment: severity-based assessment of simulated spasticity of knee joint. In: Proceedings of the 2013 IEEE/SICE International Symposium on System Integration, pp. 300–305 (2013)
4. Kikuchi, T., Oda, K., Furusho, J.: Leg-robot for demonstration of spastic movements of brain-injured patients with compact magnetorheological fluid clutch. Adv. Robot. **24**(5–6), 671–686 (2010)
5. McCrea, P.H., Eng, J.J., Hodgson, A.J.: Linear spring-damper model of the hypertonic elbow: reliability and validity. J. Neurosci. Methods **128**(1–2), 121–128 (2003)
6. Park, H.S., Kim, J., Damiano, D.L.: Development of a haptic elbow spasticity simulator (HESS) for improving accuracy and reliability of clinical assessment of spasticity. IEEE Trans. Neural Syst. Rehabil. Eng. **20**(3), 361–370 (2012)

Development of the Haptic Device for a Hepatectomy Simulator

Yu-uki Enzaki, Hiroaki Yano, Yukio Oshiro, Jaejeong Kim, Sangtae Kim, Hiroo Iwata and Nobuhiro Ohkohchi

Abstract In this study, a system for a hepatectomy simulator was developed. The system consists of 2 haptic devices, a visual display, a depth sensor, and a control PC. One of the haptic devices is used for left hand. A silicon rubber mock-up of the lobe of a liver is attached to the end effector of the haptic device. Another haptic device is provided for right hand, which is a mock-up of an ultrasonic aspirator. Then, the user can incises a virtual liver with it, and the mock-up is vibrated similar to a real ultrasonic aspirator. By using the depth sensor, the position of the haptic device is measured and transmitted to the PC. In the PC, simulation software, named Liversim, calculates the interference between the tip of the interface and the virtual liver. The virtual liver can be deformed, translated, or incised by the user. Image of the virtual liver is presented to the user. And the user can feel the reaction force like a real surgery through the haptic interfaces.

Keywords Haptic · Hepatectomy · Surgery · Simulation · Training

Y. Enzaki (✉) · H. Yano · H. Iwata
Faculty of Engineering, Information and Systems, University of Tsukuba, Tsukuba, Japan
e-mail: enzaki@vrlab.esys.tsukuba.ac.jp

H. Yano
e-mail: yano@iit.tsukuba.ac.jp

H. Iwata
e-mail: iwata@kz.tsukuba.ac.jp

Y. Oshiro · N. Ohkohchi
Faculty of Medicine, University of Tsukuba, Tsukuba, Japan
e-mail: oshiro@md.tsukuba.ac.jp

N. Ohkohchi
e-mail: nokochi3@md.tsukuba.ac.jp

J. Kim · S. Kim
Faculty of Library, Information and Media Science, University of Tsukuba, Tsukuba, Japan
e-mail: kim.jaejeong.fa@u.tsukuba.ac.jp

S. Kim
e-mail: pacman@slis.tsukuba.ac.jp

H. Kajimoto et al. (eds.), *Haptic Interaction*, Lecture Notes in Electrical Engineering 277, DOI 10.1007/978-4-431-55690-9_44

1 Introduction

The reduction of the number of surgeons and the regional health disparities has become more serious issues in recent years in Japan. The main problem is that it takes long periods to become a full-fledged surgeon. An interactive surgical training system is important for shortening the periods of mastering the surgical skill. As a related work, laparoscopic simulators such as LapSim [1] has been developed and put into practice, and laparotomy simulators have also been developed [2]. However, a simulator to touch a virtual liver by a user's bare hand has not been developed yet. Therefore, we developed a hepatectomy simulator, in which the user can touch a virtual liver by his/her own bare hand.

Figure 1 shows a future image of our surgical simulator for surgical training. The trainees watch a 3D computer graphic image of a virtual liver. Also, they can touch the image and feel the reaction force from the liver. Of course, they can cut the liver freely. The surgery is a two-handed operation. For example, a right-handed surgeon uses his/her left hand to deform or move the human organ for enhancing the visibility of the surgical field. The right hand is used for an excision with an ultrasonic aspirator. Therefore, two different types of interfaces for each hand are required for the surgical simulator. In this study, a prototype system for both hands' operation was developed.

2 System Configuration

2.1 Hardware Configuration

The prototype system consists of an interface for left hand, an interface for right hand, a visual display, a depth sensor, and a control PC (Fig. 2).

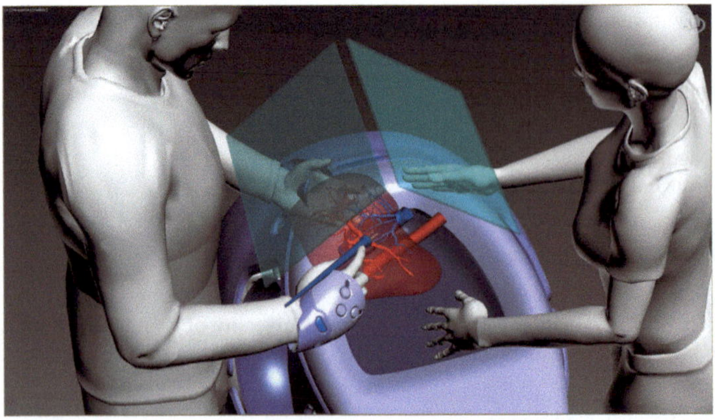

Fig. 1 Future image of the surgical simulator for 2 surgeons

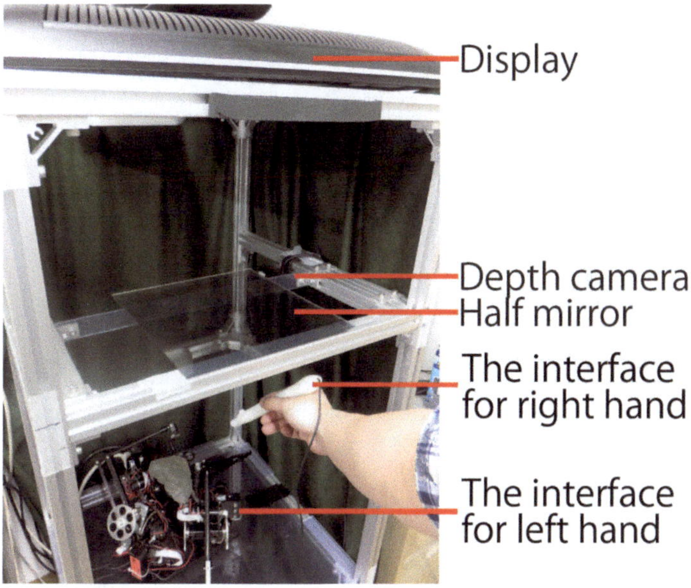

Fig. 2 System overview

The interface for left hand consists of the HapticMaster [3] and a transparent rubber mock-up of the right lobe of a liver (Fig. 3). The mock-up of the liver is attached to the end effector of the HapticMaster. The mock-up is made from silicon rubber, and its stiffness is similar to the human liver. The interface has working volume of 30 cm diameter and generates reaction force from a virtual liver to a user. The maximum reaction force is 10 N. The user can touch, deform, grasp, or move the mock-up like a real liver in a real hepatectomy by his/her own bare left hand.

Fig. 3 The interface for left hand

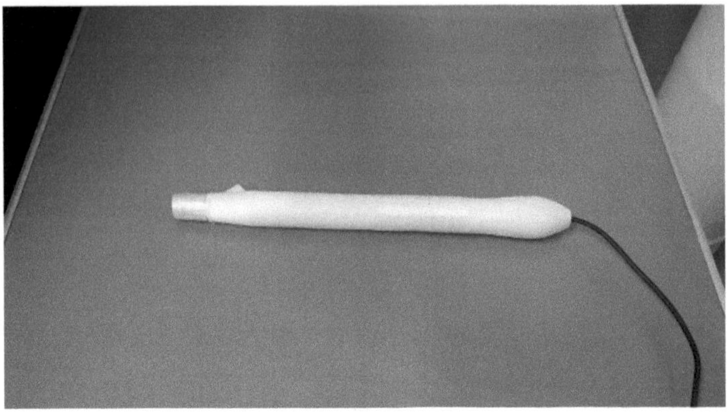

Fig. 4 The interface for right hand

Table 1 The specification of the depth camera

Range	0.15–1.0 m
FOV	74°H, 57.9°V
Resolution	320 × 240
Frame rate	60 fps
Interface	USB2.0

The interface for right hand is a mock-up of an ultrasonic surgical aspirator (Fig. 4). The mock-up is built in a vibrator, named Haptuator [4, 5], so that the user can feel the vibration like the real aspirator. At the end of the interface, a retroreflector is attached and its position and orientation are measured by the depth camera. The depth camera, DepthSense (DS325, made by SoftKinetic [6]), is used. This camera captures depth images by using a method of time of flight (TOF). The specification of the camera is shown in Table 1. The PC controls all the devices and run a surgical simulation software, named Liversim.

When the tip of the interface for right hand touches the surface of the virtual liver and a foot switch placed under the user's foot is pressed, the user can incise the virtual liver as if real incision in a real surgery.

2.2 *Software Configuration*

Liversim [7] on the control PC calculates the behavior of a virtual liver, such as deformation by the left hand, or incision with the virtual surgical aspirator. Since Liversim is basically controlled with a keyboard and a mouse, the keyboard and the mouse input are replaced with data of the haptic devices. Figure 5 shows the snapshot of deformation of a virtual liver by left hand and incision by the

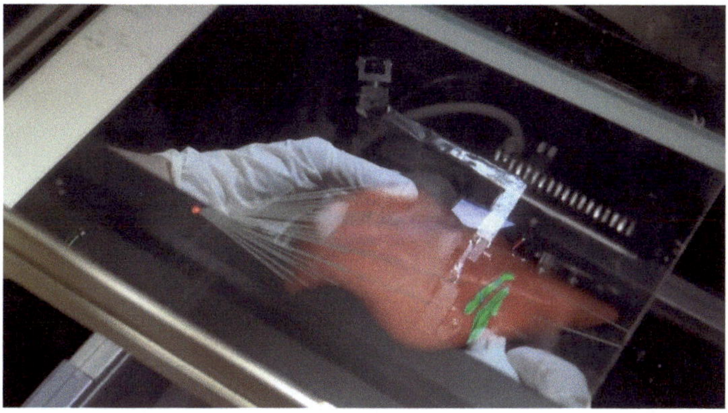

Fig. 5 Deformation and incision of the virtual liver

right hand. Since the liver is fixed by thread during liver surgery, the virtual liver is also fixed the left lobe. This system is fed back to the user the force to return the original position of the virtual liver.

3 Conclusion

In this study, a surgery simulator system was developed. As a next step, for example, we are now developing an application for rehearsal of a surgery. This application is designed for clinical surgery and education for medical students or residents. We plan to evaluate the application with medical students and residents.

References

1. Surgicalscience LapSim. http://www.surgical-science.com/
2. Suzuki, N., Hattori, A., Suzuki, S., Baur, M.P., Hirner, A., Kobayashi, S., Yamazaki, Y., Adachi, Y.: Real-time surgical simulation with haptic sensation as collaborated works between Japan and Germany. In: Medical Image Computing and Computer-Assisted Intervention—MICCAI 2001. Lecture Notes in Computer Science, Vol. 2208, pp. 1015–1021 (2001)
3. HapticMaster. http://intron.kz.tsukuba.ac.jp/hapticmaster/hapticmaster_e.html
4. TactileLabs. http://www.tactilelabs.com/
5. Yao, H.-Y., Hayward, V.: Design and analysis of a recoil-type vibrotactile transducer. J. Acoust. Soc. Am. **128**(2), 619–627 (2010)
6. SoftKinetic. http://www.softkinetic.com/
7. Oshiro, Y., Fukunaga, K., Mitani, J., Liversim, O.N.: A novel preoperative liver simulation system combined with real-time deformation of the liver. In: 23rd World Congress of the International Association of Surgeons, Gastroenterologists and Oncologists (IASGO), Bucarest, Sep 2013

Haptic Augmentation of Surgical Operation Using a Passive Hand Exoskeleton

Jun Nishida, Kei Nakai, Akira Matsushita and Kenji Suzuki

Abstract This study proposes a haptic augmentation method for surgical operations using a passive hand exoskeleton and a 4DOF link arm. The enhancements aim to allow neurosurgeons to surgerize cephalic area precisely by improving their finger movement capability. The hand exoskeleton has multiple quadric crank mechanisms with torque dampers to increase motion resolution and decrease hand tremors. The dampers are passively actuated by the action of the user. The voluntary movement of the surgeon's fingers will not be obstructed and safety would be increased. The link arm has electromagnetic brakes in each joint and is capable of supporting a surgeon's upper limb. In this study, the development and assessment of the exoskeleton and link arm are described.

Keywords Neurosurgery · Haptic augmentation · Augmented hand · Passive hand exoskeleton · Robot-assisted surgery

1 Introduction

During surgery, neurosurgeons are required to perform fine operations such as suture and neurolysis, while simultaneously understanding the three-dimensional structure of the brain. In such situations, haptic feedback is very important for

J. Nishida (✉)
School of Integrative and Global Majors, University of Tsukuba, Tsukuba, Japan
e-mail: jun@ai.iit.tsukuba.ac.jp

K. Nakai
Department of Neurosurgery, Faculty of Medicine, University of Tsukuba, Tsukuba, Japan

A. Matsushita · K. Suzuki
Center for Cybernics Research, University of Tsukuba, Tsukuba, Japan

K. Suzuki
Faculty of Engineering, Information and Systems, University of Tsukuba, Tsukuba, Japan

H. Kajimoto et al. (eds.), *Haptic Interaction*, Lecture Notes in Electrical Engineering 277, DOI 10.1007/978-4-431-55690-9_45

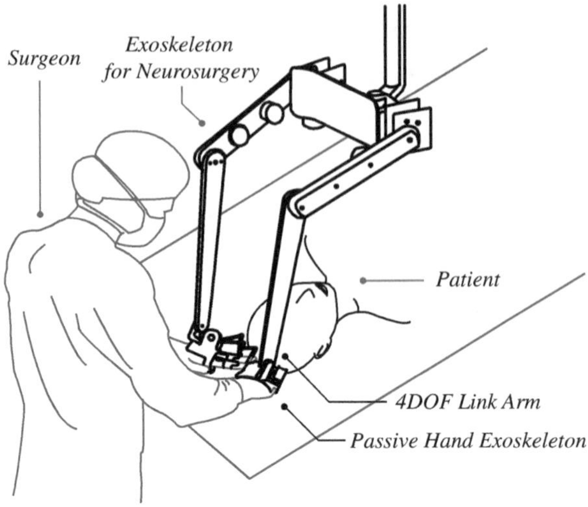

Fig. 1 Concept representation of the proposed device

understanding the condition of the patient's brain and perceiving the causal connection of the operation. To assist these operations, surgical teleoperation systems such as the da Vinci surgical system [1] have been widely used in recent years. Because the surgeon and the actual surgery field are physically separated, the haptic feedback must be estimated and displayed by sensors and actuators through an input interface. Several researches have investigated enhancement of the manipulator's operability by improving the haptic feedback systems and latency. Hannaford et al. [2] reported that a low-latency haptic feedback system allows operators to accomplish tasks faster. Kim et al. [3] reported that the latency between a manipulator and a controller affects the operability. Arata et al. [4] developed a body ownership-enhanced master–slave system by using the rubber hand illusion. However, the haptic feedback is not complete and differs from what the human body perceives.

In this research, we propose a novel assistive exoskeleton that can perform precise and augmented surgical operations without disturbing natural perception (Fig. 1). The goal of this research was to provide an augmented surgical procedure that increases finger motion resolution while decreasing hand tremors to improve reliability. The surgeons are robotically assisted by the passive hand exoskeleton, which does not introduce remote operation between the surgeon and the surgical field. The dampers that are installed in the hand exoskeleton are passively actuated by the user's action. This enables the user to perceive the actual surgical field itself. Robotic hand exoskeletons have been proposed for the use in rehabilitation [5]. However, compared to teleoperating systems, passively assisted robotic hand exoskeletons have several advantages for neurosurgery. We focused on the assistance in two areas: precise operations such as neurolysis and suturing, and securing space to facilitate an operation by enhancing the visibility of the surgical field.

2 Methodology

The proposed assistive device consists of a pair of passive hand exoskeletons and a supporting link arm that holds the hand exoskeletons and supports the surgeon's upper limb.

2.1 Passive Hand Exoskeleton

Figure 2 shows an overview of the enhancement and augmentation. The hand exoskeleton has two assistive functions: One is to stabilize and reduce tremors of the index finger by using a viscosity mechanism (torque damper), and the other is to lock the fourth and fifth fingers by using electromagnetic brakes enabling surgical spatulas to be held and used to secure the surgical space.

Operation Stabilizer To stabilize and decrease tremors of the index finger, a torque damper is attached on the back of the hand as illustrated in Fig. 3. The tip of the index finger and the torque damper are connected by a multiple quadric crank mechanism that transmits the haptic actuation through the torque damper to the tip of the finger. Although the surgical motions are minute, the angular displacement of the DIP joint (the first joint of the index finger) will be enhanced by the crank mechanism attached along the index finger. Therefore, any tremor will be decreased and the motion resolution of the finger will be increased. The motion scaling ratio M_r defined as $\frac{\Delta\theta_o}{\Delta\theta_i}$ can be calculated by using Eqs. (1)–(3). A, B, C, D are the crank length, and $\theta_1, \theta_2, \theta_3, \theta_4$ are angles illustrated in Fig. 3

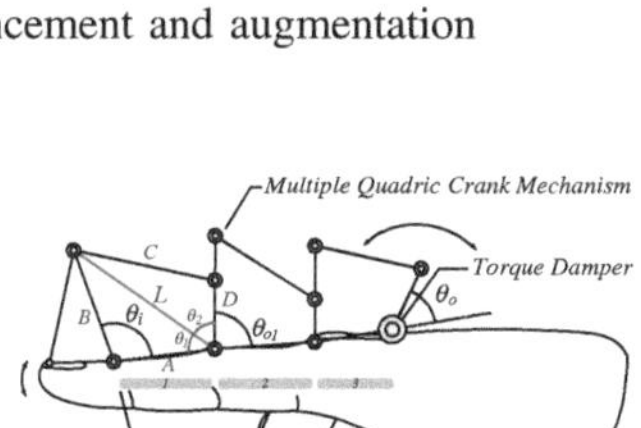

Fig. 2 Overview of the enhancement and augmentation

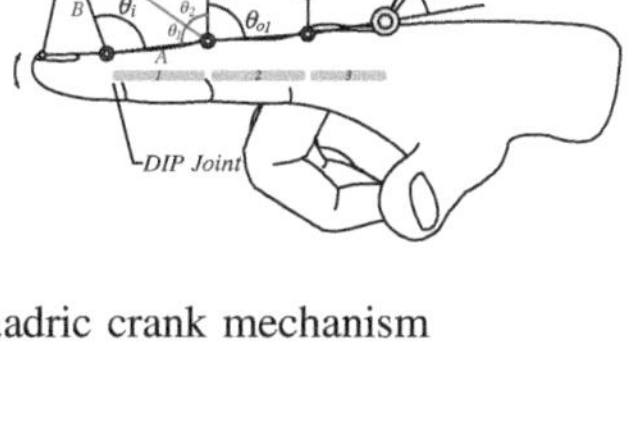

Fig. 3 Model of a multiple quadric crank mechanism

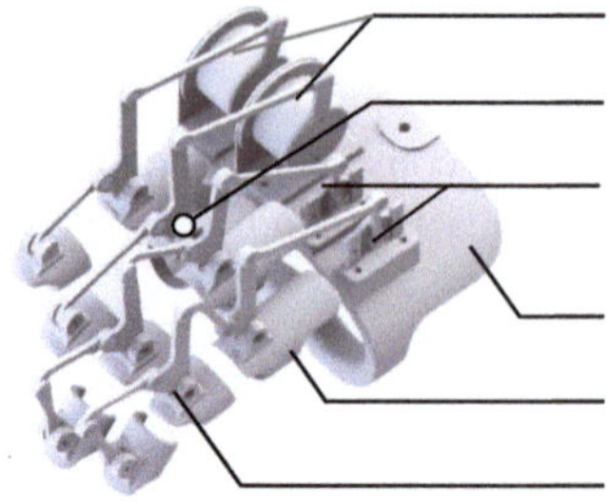

Fig. 4 Overview of the passive hand exoskeleton

$$\theta_1 = \cos^{-1}\{(A - B\cos\theta_i)/L\} \tag{1}$$

$$\theta_2 = \cos^{-1}\{(D^2 + L^2 - C^2)/2AL\} \tag{2}$$

$$\theta_{o1} = \pi - \theta_1 - \theta_2 \tag{3}$$

Locking Function for Securing Surgical Space In neurosurgery, stationary-type surgical spatulas are used to secure the surgical space in order to enhance the visibility of the surgical field and to facilitate the physical operation. In general, this is done by another surgeon who assists the lead surgeon.

We considered the use of the fourth and fifth fingers, which are not often used during surgery, and proposed to give them the additional function of assisting in space securing. Specifically, we use spatulas attached to the fourth and fifth fingers with electromagnetic brakes controlled by foot switches, to lock the finger movement at desired positions by foot switches as shown in Fig. 4. Thus, surgeons are able to expand the surgical space without requiring the support of others by using the spatulas attached to their fourth and fifth fingers.

2.2 *Upper Limb Support Link*

The support link arm is used to hold the passive hand exoskeletons and to support the surgeon's upper limb. Because it also has electromagnetic brakes in each joint, a surgeon can lock and unlock the position of the link arm at a desired position. Because the upper limb of the surgeon is supported during the surgery, hand tremors and fatigue will be reduced.

3 System Configuration

Figure 5 presents the system configuration of the proposed device. The dimensions of the hand exoskeleton were determined according to a surgeon's hand measurements, taken by a 3D shape scan device (NEC Engineering Inc., Danae 100SP).

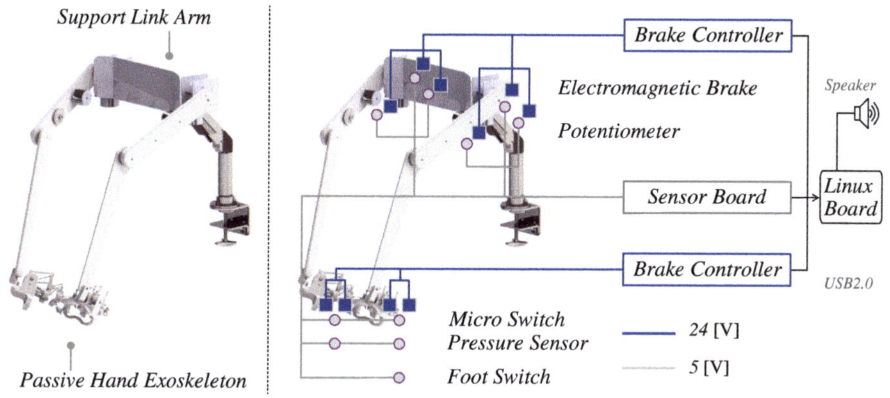

Fig. 5 System configuration of the proposed device

Each part was manufactured from a UV-hardened resin by using a 3D printer. A rotary damper (Fuji Latex Inc., FRT-G2) was used for the viscosity mechanism. The total weight of the exoskeleton is 310 g (Fig. 6).

The length of each link arm and the desired torque of the brakes were determined by the motion range simulation and evaluation of the actual position data from a surgeon's hand movements during a surgery, acquired by a motion capture system (OptiTrack Inc., V100:R2). The brakes are controlled by foot switches through relay drivers. A pressure sensor is attached below the hand exoskeleton to acquire the load that is applied by the surgeon and to sound an alert when this exceeds a safe level.

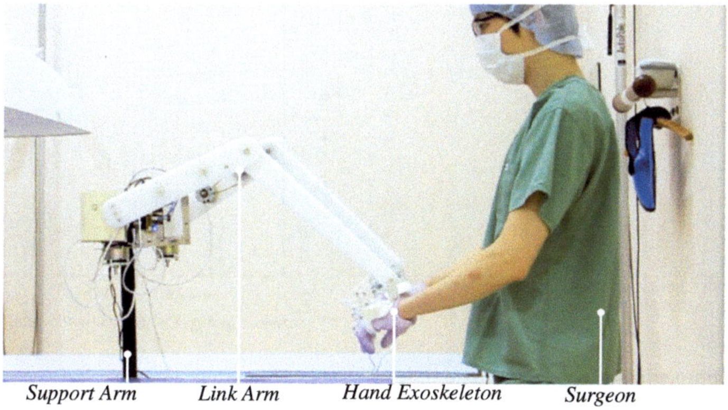

Fig. 6 Overview of the proposed device

4 Experiment

4.1 Evaluation of a Multiple Quadric Crank Mechanism

We evaluated the fundamental performance of the angular amplification mechanism, which consists of multiple quadric cranks. The length of each link was determined by simulation experiments as $(A_1, B_1, C_1, D_1), (A_2, B_2, C_2, D_2), (A_3, B_3, C_3, D_3) = (25, 15, 30, 30), (35, 15, 30, 40), (75, 20, 60, 40)(\text{mm})$.

Figure 7a shows the simulation results of θ_i and θ_o. The motion scaling ratio $M_r = 13.6$. Figure 7b shows the performance results of the actual link mechanism, where the angular change was acquired by using motion capture. The actual link mechanism achieved amplification by a factor of 12.0, which is approximately equal to the simulated result.

4.2 Hand Tremor

We evaluated the hand tremor with and without the proposed device. A subject was asked to carry out the task of holding a pair of tweezers equipped with a motion capture marker on the tip, for 60 s. Figure 7 shows a comparison of the average displacement of the tweezers' tip, which clearly indicates a reduction in tremors due to the use of the proposed device.

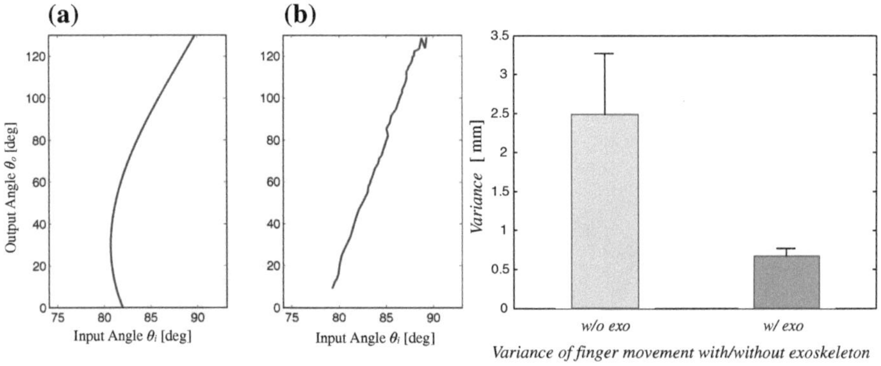

Fig. 7 Experimentation results

5 Discussion and Conclusion

The error between the link simulation and the actual experiment was due to the friction and play of each joint. Frictional spring-loaded joints may be used to improve the performance of the actual mechanism. We also verified that the locking mechanism for securing the surgical space was functioning. We believe that the use of the augmented hand for surgery would suggest a new procedure for surgeons with reduced physical and cognitive demands.

In this study, we proposed a hand exoskeleton with haptic augmentation and a support link to improve the accuracy of precise surgical operations through the suppression of finger tremors. By combining a finger locking mechanism to the exoskeleton, a surgeon has access to new functionality such as securing space without the support of others. Because the proposed haptic augmentation does not separate the surgical operation from the surgical field, it leaves the natural perception of the surgeon intact. Through experiments, we verified that the proposed system can significantly reduce finger tremors.

References

1. Guthart, G.S., et al.: The intuitive telesurgery system: overview and application. In: Proceedings of International Conference on Robotics and Automation, pp. 618–621 (2000)
2. Hannaford, B., et al.: Performance evaluation of a six-axis generalized force-reflecting teleoperator. IEEE Trans. Syst. Man Cybern. **21**(3), 620–633 (1991)
3. Kim, W.S., et al.: Force-reflection and shared compliant control in operating telemanipulators with time delay. IEEE Trans. Robot. Autom. **8**(2), 176–185 (1992)
4. Arata, J., et al.: A study on intuitive operability based on body ownership transfer for surgical assistive master-slave system. Jpn. Soc. Comput. Aided Surg. **13**(8), 174–175 (2013)
5. Matsuoka, Y., et al.: An EMG-controlled hand exoskeleton for natural pinching. J. Robot. Mechatron. **16**(5) (2004)

Haptic Virtual Reality Training Environment for Micro-robotic Cell Injection

Syafizwan Faroque, Ben Horan, Husaini Adam, Mulyoto Pangestu and Samuel Thomas

Abstract Micro-robotic cell injection is typically performed manually by a trained bio-operator, and success rates are often low. To enhance bio-operator performance during real-time cell injection, our earlier work introduced a haptically-enabled micro-robotic cell injection system. The system employed haptic virtual fixtures to provide haptic guidance according to particular performance metrics. This paper extends the work by replicating the system within a virtual reality (VR) environment for bio-operator training. Using the virtual environment, the bio-operator is able to control the virtual injection process in the same way they would with the physical haptic micro-robotic cell injection system, while benefiting from the enhanced visualisation capabilities offered by the 3D VR environment. The system is achieved using cost-effective components offering training at much lower cost than using the physical system.

Keywords Haptics · Virtual reality · Cell injection · Virtual training

S. Faroque (✉) · B. Horan · H. Adam · S. Thomas
School of Engineering, Deakin University, Geelong, Australia
e-mail: smohdfar@deakin.edu.au

B. Horan
e-mail: ben.horan@deakin.edu.au

H. Adam
e-mail: hmohdada@deakin.edu.au

S. Thomas
e-mail: sam.karukayilthomas@deakin.edu.au

M. Pangestu
Department of Obstetrics and Gynaecology, Monash University, Melbourne, Australia
e-mail: mulyoto.pangestu@monash.edu

H. Kajimoto et al. (eds.), *Haptic Interaction*, Lecture Notes
in Electrical Engineering 277, DOI 10.1007/978-4-431-55690-9_46

1 Introduction

Cell injection involves injecting material into biological cells and has been widely adopted in cell biology research, drug development, and in vitro fertilisation [1]. The process is typically performed manually by a trained bio-operator, but even so success rates are often low [2].

Micro-robotics plays a significant role in the bio-manipulation of biological cells which includes operations such as positioning, grasping and injection [3]. The micro-injection process requires approaching an immobilised cell with a micropipette, puncturing the membrane, penetrating and stopping at a suitable location inside the cytoplasm, and then depositing the material [4]. Injection success can be defined by various metrics including injection trajectory, accuracy and speed [5].

Haptic technologies interact with the human bio-operator through their haptic sensory modality and offer the ability to assist in performing cell injection. In our earlier work, a haptically-enabled system was introduced (Fig. 1, *left*) for assisting the bio-operator with certain aspects of real-time micro-robotic cell injection [6, 7]. Haptic virtual fixtures (Fig. 1, *right*) were introduced to provide the bio-operator with real-time assistance according to particular metrics. Unlike an autonomous approach to cell injection [8], the haptic virtual fixtures assist the bio-operator while they retain full control of the cell injection process.

While some other works focus on providing the bio-operator with real-time haptic feedback of cell penetration forces [9, 10], suitable sensors able to be directly employed in a laboratory environment are not yet readily accessible. As such, our physical haptic micro-robotic injection system focuses on providing haptic assistance to the bio-operator independent of the ability to measure in real-time the cell penetration/injection forces.

Aside from haptically assisting the bio-operator while performing real-time cell injection with the physical cell injection system, the approaches introduced in our earlier work can be used for training bio-operators offline in a virtual environment. Approaches employing VR and haptics for operator training have been proposed for applications such as surgical drilling [11] and minimally invasive surgery [12]. It is suggested that using a virtual replication of the physical system can enable

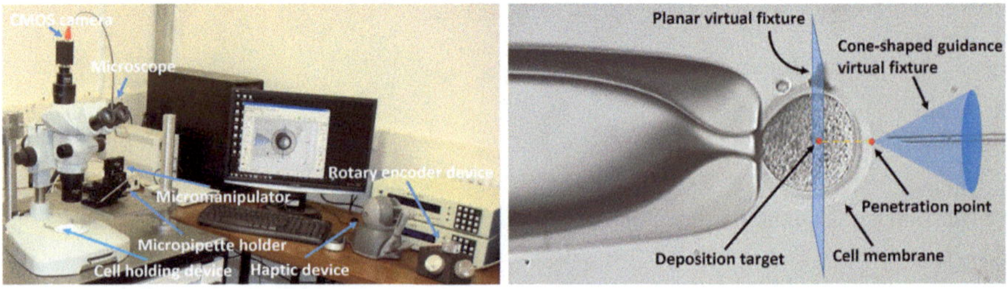

Fig. 1 Physical haptic micro-robotic cell injection system (*left*); visualisation of haptic virtual fixtures (*right*)

bio-operators to train offline in a similar manner to which they would if using the physical system. As such, the haptic virtual fixture guidance can be used both as a training aid and for real-time cell injection. It is suggested that after an appropriate amount of training has been achieved, trainee bio-operators can then transfer their skills to the physical system. This paper introduces a system which extends upon our earlier work to provide haptic virtual fixture guidance within a VR environment. Doing so can provide several benefits including low maintenance and reduced training costs.

2 Haptic VR Training Environment

The haptic VR training environment utilises the reconfigurable haptic interface (Fig. 2, *left*) introduced in our previous work [13]. The haptic interface uses two Phantom Omni [14] haptic devices and has the computer and power supply enclosed within the base of the system.

In order to represent the injection of a cell within a virtual environment, a suitable model able to provide realistic haptic and visual responses to penetration is required. Models based on boundary element modelling have been proposed for the estimation of cell interaction forces [15]. In our recent work, a particle-based cell model able to provide realistic cell responses to interaction within a virtual environment was introduced [16]. Using the model, the virtual cell visually deforms in response to penetration as well providing an estimate of the interaction forces. These forces can be haptically displayed to the bio-operator and correspond to those which may be measured by force sensors integrated to the physical system once they become readily available.

To perform cell injection using the VR training system, the bio-operator holds the stylus of the haptic device and is able to control the micro-robot and utilise the haptic virtual fixtures. The virtual deformable cell, virtual fixtures, micro-robot, etc., can be viewed in the environment. Unlike the physical cell injection system, the

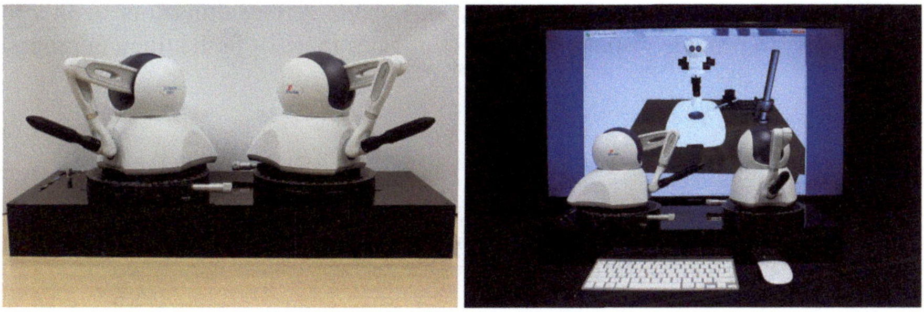

Fig. 2 Reconfigurable multipurpose haptic interface [13] (*left*); haptic VR training environment for cell injection (*right*)

operator has a 3D view of all system components as well as being able to zoom into focus on areas of interest.

The movement of the injection micropipette is mapped to the stylus of the haptic device in the same way as our earlier introduced haptic micro-robotic cell injection system. As the bio-operator commands the micropipette to approach the cell, a conical potential field virtual fixture encourages them to follow an optimised trajectory to the penetration point on the cell membrane. Once the micropipette's tip has reached the penetration point, and the bio-operator attempts to pierce the cell membrane, they will feel the simulated reaction force, followed by a sudden drop in force indicating rupture and penetration of the cell membrane. The bio-operator then needs to guide the micropipette along a straight path towards the deposition location in the direction of the micropipette's longitudinal axis. An axial virtual fixture then guides the bio-operator's movement along this path. In order to prevent the bio-operator from overshooting the deposition location, which can jeopardise the process, a planar virtual fixture is provided, attempting to prevent the bio-operator's control of the micropipette from passing the target location. Once the micropipette's tip reaches the deposition location, the virtual cell becomes inactive and the system can be reset to perform a new trial. It is important to remember that the virtual fixtures only provide haptic guidance, and given the maximum amount of force displayable by the haptic device, the bio-operator remains in full control of the injection process. The virtual fixtures can be used with or without visual overlay and can be turned off based on the trainee bio-operator's preference.

3 Conclusion and Future Work

This paper presents a haptic VR training environment modelled on our physical haptically-enabled cell injection system. Using the training environment, it is suggested that bio-operators can train offline in a similar way to which they would with the real-world system and then after receiving adequate training, they can transfer their skills to the physical system. The VR system is achieved using cost-effective components offering training at much lower cost than could be achieved using a physical system. Directions for future work include further utilisation of the VR functionality to provide visual instructions and cues to assist in the training process.

References

1. Kuncova, J., Kallio, P.: Challenges in capillary pressure microinjection. In: 26th IEEE Annual International Conference on Engineering in Medicine and Biology Society, pp. 4998–5001 (2004)
2. Pillarisetti, A., Pekarev, M., Brooks, A.D., Desai, J.P.: Evaluating the role of force feedback for biomanipulation tasks. In: Symposium on Haptic Interfaces for Virtual Environment and Teleoperator Systems, pp. 11–18 (2006)

3. Yu, S., Nelson, B.J.: Biological cell injection using an autonomous microrobotic system. Int. J. Rob. Res. **21**, 861–868 (2002)
4. Ghanbari, A., Horan, B., Nahavandi, S., Chen, X., Wang, W.: Haptic microrobotic cell injection system. IEEE Syst. J. **8**, 1 (2012)
5. Kimura, Y., Yanagimachi, R.: Intracytoplasmic sperm injection in the mouse. Biol. Reprod. **52**, 709–720 (1995)
6. Ghanbari, A., Abdi, H., Horan, B., Nahavandi, S., Chen, X., Wang, W.: Haptic guidance for microrobotic intracellular injection. In: 3rd IEEE RAS and EMBS International Conference on Biomedical Robotics and Biomechatronics, pp. 162–167 (2010)
7. Ghanbari, A., Chen, X., Wang, W., Horan, B., Abdi, H., Nahavandi, S.: Haptic microrobotic intracellular injection assistance using virtual fixtures. In: 11th International Conference on Control, Automation, Robotics and Vision, pp. 781–786 (2010)
8. Wang, W.H., Liu, X.Y., Sun, Y.: High-throughput automated injection of individual biological cells. IEEE Trans. Autom. Sci. Eng. **6**(2), 209–219 (2009)
9. Pillarisetti, A.: Force feedback interface for cell injection. In: Proceedings of 1st Joint Eurohaptics Conference and Symposium on Haptic Interfaces for Virtual Environment and Teleoperator Systems, pp. 391–400 (2005)
10. Cho, S.Y., Shim, J.H.: A new micro biological cell injection system. In: Proceedings of the IEEE/RSJ IROS International Conference on Intelligent Robots and Systems, pp. 1642–1647 (2004)
11. Sewell, C., Blevins, N.H., Peddamatham, S., Tan, H.: The effect of virtual haptic training on real surgical drilling proficiency. In: 2nd Joint Eurohaptics Conference and Symposium on Haptic Interfaces for Virtual Environment and Teleoperator Systems, pp. 601–603 (2007)
12. Basdogan, C., De, S., Kim, J., Muniyandi, M., Kim, H., Srinivasan, M.A.: Haptics in minimally invasive surgical simulation and training. IEEE Comput. Graphics Appl. **24**, 56–64 (2004)
13. Horan, B., Faroque, S., Isaksson, M., Ang, Q.: Reconfigurable multipurpose haptic interface. In: Auvray, M., Duriez, C. (eds.) Haptics: Neuroscience, Devices, Modeling, and Applications, Part II. LNCS, vol. 8619, pp. 432–434. Springer, Heidelberg (2014)
14. Geomagic, http://geomagic.com/
15. Kim, J., Janabi-Sharifi, F., Kim, J.: Haptic feedback based on physically based modeling for cellular manipulations systems. In: Ferre, M. (ed.) Haptics: Perception, Devices and Scenarios. LNCS, vol. 5024, pp. 661–667. Springer, Heidelberg (2008)
16. Horan, B., Lowe, D., Ang, Q-Z., Asgari, M., Ghanbari, A., Nahavandi, S.: Virtual haptic cell model for operator training. In: Proceedings of the 15th International Conference on Mechatronics Technology, pp. 1–5 (2011)

Part VI
VR, Telepresence and Multimedia

FeelCraft: User-Crafted Tactile Content

Oliver Schneider, Siyan Zhao and Ali Israr

Abstract Despite ongoing research into delivering haptic content, users still have no accessible way to add haptics to their experiences. Lack of haptic media infrastructure, few libraries of haptic content, and individual differences all provide barriers to creating mainstream haptics. In this paper, we present an architecture that supports generation of haptic content, haptic content repositories, and customization of haptic experiences. We introduce *FeelCraft*, a software plugin that monitors activities in media and associates them with expressive tactile patterns known as *feel effects*. The FeelCraft plugin allows end users to quickly generate haptic effects, associate them to events in the media, play them back for testing, save them, share them, and/or broadcast them to other users to feel the same haptic experience. The FeelCraft architecture supports both existing and future media content, and can be applied to a wide range of social, educational, and assistive applications.

Keywords Feel effect · Haptic authoring · Haptic broadcasting · Haptic media

1 Introduction

In recent years, haptic feedback has shown promise to enhance user experience in movies, games, rides, virtual simulations, and social and educational media [1–3]. However, current mainstream media has yet to use the richness of haptic modality within its content. The lack of haptic authoring tools, production infrastructure,

O. Schneider (✉) · S. Zhao · A. Israr
Disney Research Pittsburgh, 4720 Forbes Avenue, Pittsburgh, PA, USA
e-mail: oschneid@cs.ubc.ca

S. Zhao
e-mail: siyan.zhao@disneyresearch.com

A. Israr
e-mail: israr@disneyresearch.com

O. Schneider
University of British Columbia, 201-2366 Main Mall, Vancouver, BC, Canada

H. Kajimoto et al. (eds.), *Haptic Interaction*, Lecture Notes
in Electrical Engineering 277, DOI 10.1007/978-4-431-55690-9_47

standardized playback protocols, and skilled and trained workers has contributed to the difficulty of integrating haptic content with accompanying media. To reduce the gap between haptics and mainstream communication, haptic feedback must be expressive, coherent, and synchronized with the content of the media, and also meet user expectations. We believe that allowing end users to access, customize, and share haptic media will create an intimate, engaging, and personalized experience, and proliferate the use of haptics.

In this paper, we propose and implement an architecture that channels media content to dynamic and expressive tactile sensations. We introduce FeelCraft, a media plugin that monitors events and activities in the media, and associates them to user-defined haptic content in a seamless, structured way. The FeelCraft plugin allows novice users to generate, recall, save, and share haptic content, and play and broadcast them to other users to feel the same haptic experience, without having any skill in haptic content generation. In the current implementation, we concentrate on the vibrotactile array as the source of sensation and the back as the surface for stimulation; however, the FeelCraft architecture can be easily adapted for other haptic feedback modalities.

We begin by presenting relevant background work related to haptic media infrastructure. We then present the FeelCraft architecture and describe in detail each component of the system. Finally, we conclude the paper with our envisioned application ecosystem.

2 Related Work

Infrastructures to integrate haptic feedback in media have been primarily derived from media type and user interactions associated with the content. These infrastructures fall into two categories: event triggers and direct mappings.

In the **event triggers** scheme, haptic information is embedded in the media and played back using predefined protocols [4]. A common example is a video game controller that rumbles on predefined triggers embedded in the games. This technique is complex and requires a proper production infrastructure that is expensive. Another disadvantage of this technique is that media designers would require access to tools and libraries for creating expressive haptic effects, similar to the libraries for visual and sound effects [5].

Direct mappings use cues from existing media and directly maps them to haptic effects. For example, a typical way to enhance movies and other visual content with haptics is to monitor activity in a video feed [6] and map these activities to haptic transducers arranged along the seat [1, 7]. This way, movements in a visual scene are mapped to gross motion collocated with events seen in 4D movies and rides (www.d-box.com). Similarly, sound has been used to derive haptic cues for video games and music [8, 9]. For example, Buttkicker technology (Guitammer, USA) shakes the entire seat using low-pass filtered sound. The Vybe Haptic Gaming Pad (Comfort Research) divides sound into three bands and maps them to transducers located in the seat and back. The advantage of the direct mapping scheme is that no change is required in the current media production process. However, each technique is limited to its media type.

Tactile icon libraries have been developed independent of media types. Immersion Corp. (www.immersion.com) has devised a library that could be used by designers, but it is limited to single actuators. Other researchers have proposed libraries and authoring tools [10, 11] along with MPEG-V standardization protocols, but are limited to in-lab use and specific hardware [12–14]. Furthermore, customization and personalization have been found to be important to end users [15, 16], but most libraries do not support them. Recently, a library of feel effects (FEs, defined in [17]) shows a promising approach to creating an engaging library. Of course, libraries require an infrastructure to connect to media before they can be widely applied.

3 FeelCraft Plugin and Architecture

A FeelCraft plugin maps media to haptic sensations in a modular fashion, supporting arbitrary media types and output devices. By using a FeelCraft plugin, users can:

- link existing and new media to the haptic feedback technology,
- use an FE library to find appropriate semantically defined effects,
- author, customize, and share a common, evolving repository of FEs, and
- play and broadcast haptic experiences to one or more user(s).

A pictorial description of the FeelCraft architecture is shown in Fig. 1 architecture. The conceptual framework of FeelCraft revolves around the FE library

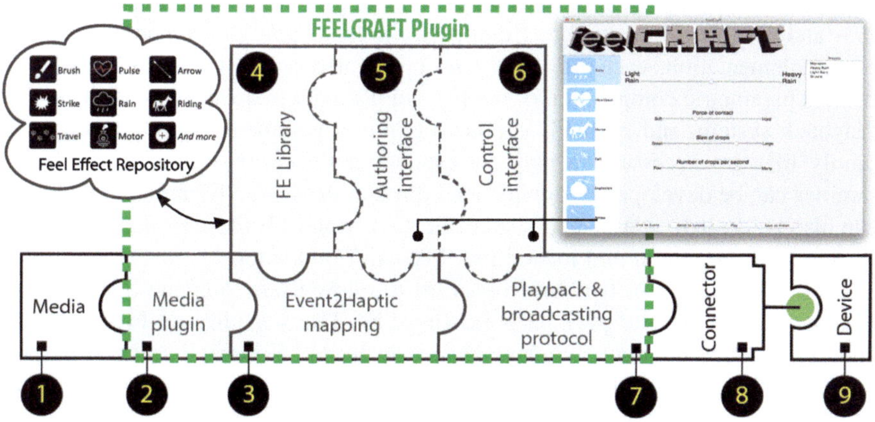

Fig. 1 FeelCraft architecture. The FeelCraft plugin is highlighted in *green*. The FE library can connect to shared feel effect repositories to download or upload new FEs. A screenshot of our combined authoring and control interface is on the *right*

introduced in [17]. The FE library provides a structured and semantically correct association of media events with haptic feedback. By using the authoring interface to tailor FE parameters, a repository of FEs can remain general while being used for unique, engaging, and suitable sensations for different media. The playback system, authoring and control interface, Event2Haptic mappings, and media plugin support seamless flow of the media content to the haptic feedback hardware.

Media (**1**) can be entertaining, such as video games, movies, and music, or social and educational. The media can also be general user activity or embedded events in applications. In our implementation, we use the popular sandbox indie game Minecraft (https://minecraft.net).

Media Plugin (**2**) is a software plugin that communicates with the media and outputs events and activities. This plugin can be as simple as receiving messages from the media or as complicated as extracting events and activities from a sound stream. With existing media, common plugin systems are automatic capture of semantic content from video frames [6], camera angles [1], or sounds [8, 9], or the interception of input devices (such as game controllers or keyboard events). We use a CraftBukkit Minecraft server modification to capture in-game events.

Event2Haptic (**3**) mappings associate events to FEs, which are designed, tuned, and approved by users using the FE library. This critical component links the media plugin's output to the haptic playback system. Currently, six FEs are triggered by six recurring in-game events: the presence of rain, low player health, movement on horse, strike from a projectile, in-game explosions, and player falls. Our implementation provides the option to store this mapping directly in the source code, or in a text-based JavaScript Object Notation (JSON) file.

FE Library (**4**) is a collection of FEs. A key feature of an FE is that it correlates the semantic interpretation of an event with the parametric composition of the sensation in terms of physical variables, such as intensity, duration, and temporal onsets [17]. Each FE is associated with a family, and semantically, similar FEs are associated with the same family. For example, the Rain family contains FEs of *light rain* and *heavy rain*; as well that that of *sprinkle*, *drizzle*, *downpour,* and *rain*. In our implementation, each FE family is represented as a Python source file that defines parametric composition of the FE and playback sequences for the FeelCraft Playback system, and each FE is coded as preset parameters in a JSON file. FE family files are necessary to play corresponding FEs in the family, and new FE families can be developed or downloaded through the shared FE repository. The FE can also be created, stored, and shared. FE family and FE files are stored in a local directory of the plugin and loaded into FeelCraft on startup.

Authoring and Control Interfaces (**5, 6**) allow users to create and save new FEs and tune, edit, and play back existing FEs. Users modify an FE by varying sliders labeled as common language phrases instead of parameters such as duration and intensity (Fig. 1). Therefore, users can design and alter FEs by only using the semantic logic defining the event. The interface also allows users to map game events to new FEs and broadcast to other users, supporting a What-You-Feel-Is-What-I-Feel (WYFIWIF) interface [15].

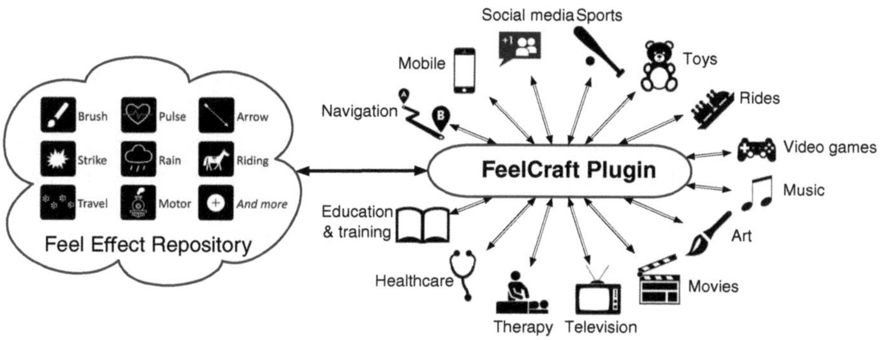

Fig. 2 Application ecosystem for FeelCraft and an FE repository

Playback and Communication Protocols (7) render FEs using the structure defined in FE family files and outputs them through a communication method (8) to one or more devices (9). Our implementation includes an API controlling the commercially available Vybe Haptic Gaming Pad via USB.

4 Application Ecosystem

FeelCraft plugins are designed to make haptics accessible to end users using existing media and technology. For example, a user may want to assign a custom vibration to a friend's phone number, or add haptics to a game. In this case, a user would download a FeelCraft plugin for their device, browse FEs on an online feel repository, and download FE families they prefer. Once downloaded, the FeelCraft authoring interface allows for customization, as a rain FE for one video game may not quite suit another game. The user could create a new FE for their specific application, and once they were happy with it, upload their custom FE for others to use. If the user wanted to show a friend their FE, they could use the playback system to drive output to multiple devices, or export the FE to a file and send it to them later. Figure 2 illustrates this ecosystem with application areas. Just like the Noun Project for visual icons (http://thenounproject.com) and downloadable sound effect libraries, we envision online repositories of FEs that can be continually expanded with new FEs by users. Our current FE repository includes six original families described in [17] and an additional four new families: *Ride*, *Explosion*, *Fall,* and *Arrow.*

5 Conclusion

In this paper, we have presented FeelCraft, a software plugin that allows users to author, customize, share, and broadcast dynamic tactile experiences with media and user activities. The plugin uses FEs, semantically structured haptic patterns.

We integrated the FeelCraft plugin with a popular sandbox game, MineCraft. Users can associate six events in the game to corresponding FE, and modify and broadcast to other users to share the haptic experience. The newly authored FEs are saved and shared with other users for communal use via an online FE repository. The proposed plugin can also be used with a wide range of entertaining, social, and educational media. Future work includes expanding the FE repository, networked communication and sharing, and supporting output to different haptic device types while maintaining semantic meaning, connecting end users to haptics in an even more accessible manner.

References

1. Danieau, F., Fleureau, J., Guillotel, P., Mollet, N., Christie, M., Lecuyer, A.: Toward haptic cinematography: enhancing movie experiences with camera-based haptic effects. IEEE Multimedia **21**(2), 11–21 (2014)
2. Farley, H., Steel, C.: A quest for the holy grail: tactile precision, natural movement and haptic feedback in 3D virtual spaces. In: ASCILITE 2009, pp. 285–295. Australasian Society for Computers in Learning in Tertiary Education (ASCILITE), Dec 2009
3. Israr, A., Poupyrev, I., Ioffreda, C., Cox, J., Gouveia, N., Bowles, H., Brakis, A., Knight, B., Mitchell, K., Williams, T.: Surround haptics: sending shivers down your spine. In: ACM SIGGRAPH Emerging Technologies, Vancouver, Canada, Aug 2011
4. Modhrain, S.O., Oakley, I.: Touch TV: adding Feeling to broadcast media (2001)
5. Cha, J., Ho, Y.S., Kim, Y., Ryu, J., Oakley, I.: A framework for haptic broadcasting. IEEE Multimedia **16**(3), 16–27 (2009)
6. Kim, M., Lee, S., Choi, S.: Saliency-driven tactile effect authoring for real-time visuotactile feedback. In: Haptics: Perception, Devices, Mobility, and Communication, vol. 7282, pp. 258–269, LNCS (2012)
7. Bach-Y Rita, P., Collins, C.C., Saunders, F.A., White, B., Scadden, L.: Vision substitution by tactile image projection. Nature **221**(5184), 963–964 (1969)
8. Chang, A., O'Sullivan, C.: Audio-haptic feedback in mobile phones. In: CHI '05 extended abstracts, pp. 1264–1267, Apr 2005
9. Lee, J., Choi, S.: Real-time perception-level translation from audio signals to vibrotactile effects. In: CHI '13, pp. 2567–2576, Paris, France, Apr 2013
10. Enriquez, M., MacLean, K.: The hapticon editor: a tool in support of haptic communication research. In: Haptics Symposium (HAPTICS '03), pp. 356–362, IEEE Computer Society (2003)
11. Swindells, C., Pietarinen, S., Viitanen, A.: Medium fidelity rapid prototyping of vibrotactile haptic, audio and video effects. In: Haptics Symposium (HAPTICS '14), pp. 515–521, IEEE, Feb 2014
12. Gao, Y., Osman, H.A., El Saddik, A.: MPEG-V based web haptic authoring tool. In: IEEE International Symposium on Haptic Audio Visual Environments and Games (HAVE), pp. 87–91, IEEE, Oct 2013
13. Kim, J., Lee, C.G., Kim, Y., Ryu, J.: Construction of a haptic-enabled broadcasting system based on the MPEG-V standard. Sig. Process. Image Commun. **28**(2), 151–161 (2013)
14. Waltl, M., Rainer, B., Timmerer, C., Hellwagner, H.: A toolset for the authoring, simulation, and rendering of sensory experiences. In: Multimedia (MM '12), pp. 1469–1472, Oct 2012
15. Schneider, O.S., MacLean, K.E.: Improvising design with a haptic instrument. In: Haptics Symposium (HAPTICS '14), Houston, USA (2014)

16. Seifi, H., Anthonypillai, C., MacLean, K.E.: End-user customization of affective tactile messages: a qualitative examination of tool parameters. In: Haptics Symposium (HAPTICS '14), pp. 251–256, IEEE, Feb 2014
17. Israr, A., Zhao, S., Schwalje, K., Klatzky, R., Lehman, J.: Feel effects: enriching storytelling with haptic feedback. ACM Trans. Appl. Percept. 17 (2014)

Development of Ball Game Defense Robot Based on Physical Properties and Motion of Human

Kosuke Sato, Yuki Hashimoto, Hiroaki Yano and Hiroo Iwata

Abstract The ball game robot has attracted considerable research attention in recent years. This paper describes the research and development of a ball game robot based on human physical properties and defensive motion. In particular, we focus on reproducing defensive motion. We will develop a defense robot that can be used in a training field and aim at building an interactive system. Considering the problem of strength and stability control, we attached weight to important operations. Because the arms and hands of the robot were built to imitate human physical properties, the robot can bounce a ball realistically. We conducted a quality assessment experiment to determine whether the robot operations indeed imitate a defensive motion and verified the validity of the experiment.

Keywords Virtual reality · Human interface · Robotics · Human robot interaction · Motion capture

K. Sato (✉)
School of Integrative and Global Majors, University of Tsukuba,
1-1-1 Tennoudai, Tsukuba-shi, Ibaraki 305-8573, Japan
e-mail: kosuke@vrlab.esys.tsukuba.ac.jp

Y. Hashimoto · H. Yano · H. Iwata
Faculty of Engineering, Information and Systems, University of Tsukuba,
1-1-1 Tennoudai, Tsukuba-Shi, Ibaraki 305-8573, Japan
e-mail: hashimoto@iit.tsukuba.ac.jp

H. Yano
e-mail: yano@iit.tsukuba.ac.jp

H. Iwata
e-mail: iwata@kz.tsukuba.ac.jp

H. Kajimoto et al. (eds.), *Haptic Interaction*, Lecture Notes
in Electrical Engineering 277, DOI 10.1007/978-4-431-55690-9_48

1 Introduction

Recently, research on virtual reality has been carried out for various fields, such as industrial, medical, and entertainment. It is also expected for virtual reality to be used in sports. In this study, we focus on ball game sports and propose a defense robot to use in training. In order to develop one's game-related judgment (ball handling in accordance with movement), a practice partner is always required. However, opportunities for performing such practice are generally rare, and scene reproducibility is low.

In this study, we proposed to use a defense robot for use in training. In the training with the robot, if a ball moves to the working volume of the robot's arm, the robot intercepts the passed ball. Since the speed of the robot motion is as same as that of human being and the response of the robot can control arbitrarily, the ball interception rate can be controlled. Therefore, the user can learn how to pass the ball correctly through trial and error with the system.

When designing a ball game defense robot, providing visual and somatosensory feedback to an offense player is an important consideration. Because offense players throw the ball to other players or a basket, they avoid colliding with defense players. To realize this, the offense players have to know the situation of the defense players. The situation is mostly acquired by using visual information such as defender's position and posture, audio information such as sound of footstep, and somatosensory information such as body contact. In this study, we focused on visual and somatosensory feedback methods. In order to realize visual feedback effectively, the mechanism of the robot's body has to resemble human being. And the robot should have ability to move like human act. Besides, with regard to the somatosensory feedback, an offense player can grasp correlation among himself/ herself, the ball through contact information. In this study, we focused on the contact information between the ball and the arm of the defense player for safety reason.

In view of control stabilization and strength, and readiness for being called the best part of sports, we do not propose a humanoid robot that mimics the shape of the human body. Therefore, we focused on the movement that is important for ball game sports and designed a robot whose function is to execute the same performance of human being.

2 System Configurations

2.1 Hardware Configurations

We focused on the game of basketball, and more specifically, on the shooting and zone defense actions executed in this game to perform typical operations of ball game sports. An analysis of the defender's movement reveals that such movement is roughly divided into jumps, use of arms, and translational motion. As the first step of this study, it is considered that the arm motion is most important motion. Therefore, 2 DOF arms were designed. Mechanisms of the upper arm and the forearm of both arms were imitated by using the four-link mechanisms (Figs. 1 and 2).

Fig. 1 Mechanism

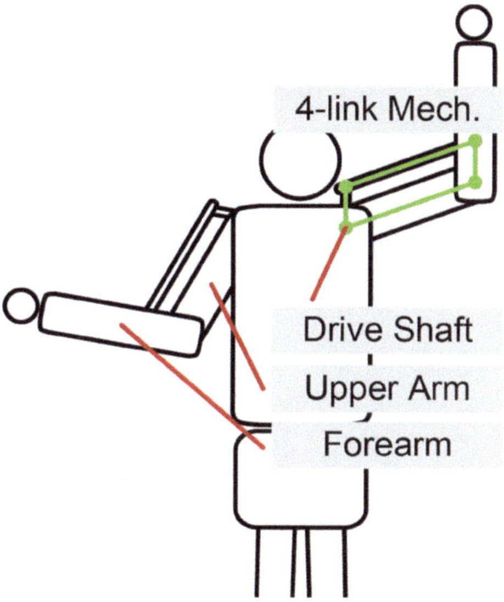

Fig. 2 Overview

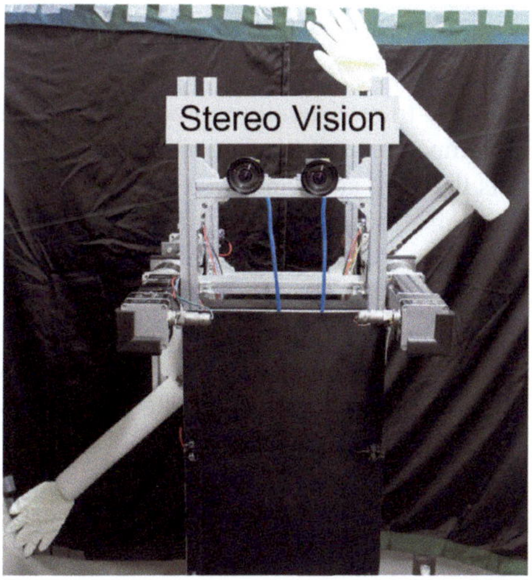

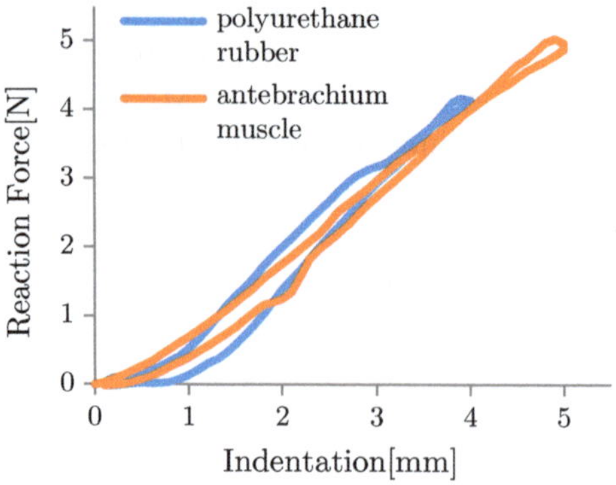

Fig. 3 Elasticity comparison

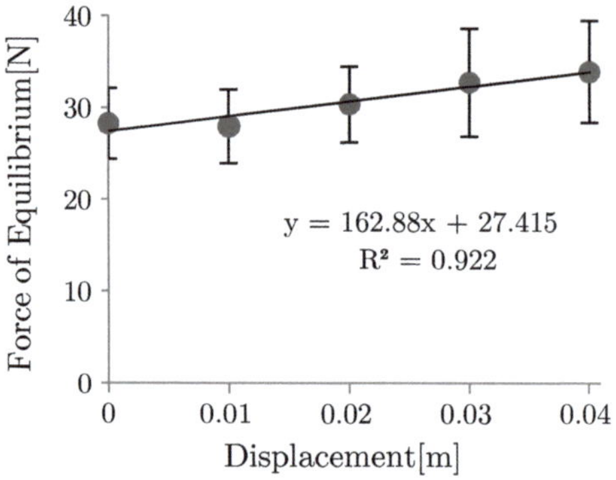

Fig. 4 Finger impedance

2.2 *Physical Properties*

In order to realistically reproduce the behavior and rebound of a ball, we measured the impedance of the fingers and muscle elasticity of the human arm when the ball hits the arms and hands of a defense player. With regard to the arm, we measured its elasticity using an elastic measuring instrument composed of a load cell (Kyowa Elec. Inst. Co. LMA-A-5N) and a linear actuator (Oriental Motor Co. ELCM2F10K) [1] (Fig. 3). By sampling a urethane rubber that has an equivalent elasticity basis, we created an arm model. With regard to the hand, we measured finger impedance using a force gauge (Tech-Jam Co. ZP-500N). When pulling a subject's finger toward the back of the hand, we compelled the subject to oppose this force (Fig. 4).

2.3 *Ball Tracking*

Measuring ball position through 3D reconstruction using stereo vision camera (Point Grey Co. GS3-U3-41C6C-C, 90 FPS) is done in order to perform an

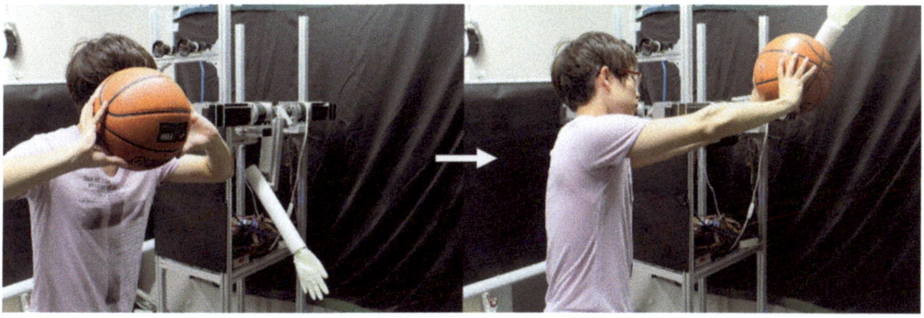

Fig. 5 The robot prevents the path course (*left*) and presents the reaction force (*right*)

operation corresponding to the movement of an offense player. Because the ball used in basketball is a single color, we utilized the stereomatching method using hue information.

To influence visual and somatosensory feedback for an offense player, it is necessary to simulate the behavior of an actual defense player. 3D position of the ball is measured by using a stereovision technique. The robot can follow the ball motion in real time and prevent the path course. If the ball moves to the working volume of the robot's arm, the robot intercepts the ball (Fig. 5). When the user is about to shoot a basket and the robot touches the ball, the user can feel the reaction force from the robot.

3 Conclusion

We limited the design and manufacture of our proposed mechanism to significant movements of real defense players. Because the arms and hands of the robot were built to imitate human physical properties, the robot can bounce a ball realistically. From now on, we will consider about flexibility extension of jumps and translational motion. In connection with it, there is a need to examine not only by interactive arm, but also with respect to body contact with the whole body of the robot.

In the future, our proposed defense robot will be used in a practical basketball game; therefore, we will verify its effectiveness as a defense robot.

Reference

1. Kimura, M., Yamamoto, M.: Comparison of side step speed toward the right and left direction in basketball player. Bulletin **27**(1), Tokyo Polytechnic University, Faculty of Engineering (2004)

Development of Handshake Gadget and Exhibition in Niconico Chokaigi

Takanori Miyoshi, Yuki Ueno, Kouki Kawase, Yusaku Matsuda, Yuya Ogawa, Kento Takemori and Kazuhiko Terashima

Abstract We developed a handshake gadget that allows people in remote locations to feel mutual force and movement and exhibited it at the 2014 Niconico Chokaigi in April. The Niconico Chokaigi is a very large exhibition in Japan; 125,000 persons attended, and more than 7 million people viewed the movie via the Internet. Our handshake gadget connected individuals in Japan and Taiwan and was experienced by a number of people, including the Japanese Prime Minister and popular characters such as Funasshi and Mario. An international questionnaire administered to those trying the device found that more than 70 % expressed excitement for this gadget, while more than 65 % could feel mutual force and motion.

Keywords Tele-operation · Handshake · Bilateral tele-control · Tele-existence · Intercontinental network experiment

1 Introduction

The handshake is one of the most popular forms of nonverbal communication. There have been various proposals as to how it might be made into a communication means over a long distance. In 1997, Ouchi and Hashimoto [1] proposed a "Handshake Telephone System" to communicate both voice and force. A similar robot was developed by Handshake Interactive Technologies, Inc., in 2003 [2], using Niemeyer's algorithm [3]. These systems exercised only an arm, however. In order to achieve a more realistic sensation, both [4, 5] also proposed a method for moving a finger, although these researchers did not consider the instability dependent on communication latency. We previously reported an example that considered both finger motion and influence by communication latency [6].

T. Miyoshi (✉) · Y. Ueno · K. Kawase · Y. Matsuda · Y. Ogawa · K. Takemori · K. Terashima
Toyohashi University of Technology, 1-1 Tempk-cho, Toyohashi, Japan
e-mail: miyoshi@tut.jp
URL: http://www.syscon.tut.ac.jp/

H. Kajimoto et al. (eds.), *Haptic Interaction*, Lecture Notes in Electrical Engineering 277, DOI 10.1007/978-4-431-55690-9_49

The purpose of this research is not to study the complicated control such as above-mentioned research, but to propose a handshake gadget that anyone can produce using simple parts, easily obtained online, and to exhibit it through an intercontinental network connection. Therefore, we used the popular haptic device Falcon [7], adding a force sensor and a sufficiently robust tele-control algorithm. That means that anyone can make the gadget and experience it anytime, anywhere. Now, tele-existence will no longer belong only to specialists.

2 Structure of System

2.1 Hardware Setup

As shown in Fig. 1, the Falcon, with the attached mannequin hands and note PCs, are set up in Taiwan and Japan. On the Taiwanese side, the 3-axis force sensor is also assembled between a Falcon and mannequin hand, and a note PC can measure the human's operating force through a USB. The Falcon and PC are connected by USB, and the PCs by the Internet. Thus, the hardware is very simple.

Subjects grasp the mannequin hand and give it a shaking motion up/down, left/right, and back/forth. The force on Taiwanese side was then transmits to the Japanese side, and the Japanese Falcon moves in accordance with Taiwanese subject's force. In the same manner, Japanese motion is transmitted to the Taiwanese side, and the Taiwan Falcon moves in accordance with Japanese motion. When each performs opposite movements, resistance can be felt. On the other hand, if subjects' movements are synchronized, attraction force can be felt. Thus, each subject experiences mutual force and movement mutually.

2.2 Software Setup

Figure 2 shows the control block diagram. T_1 and T_2 mean the communication latency of the Internet between Taiwan and Japan. Although based on wave variable control, our algorithm is more robust than [3]. That means, even if there is some packet loss and/or perturbation of communication latency, the position of the

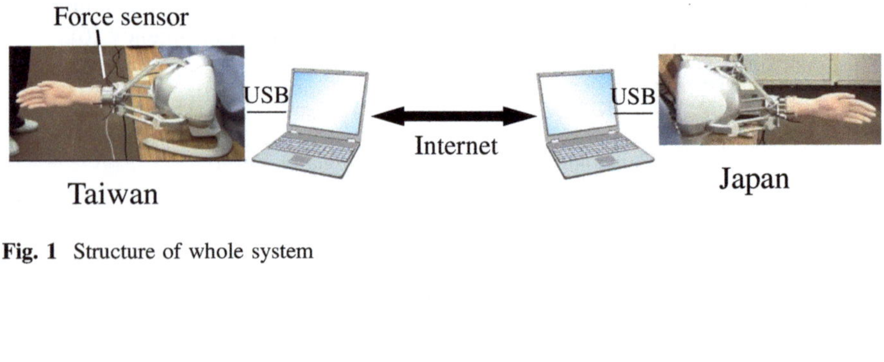

Fig. 1 Structure of whole system

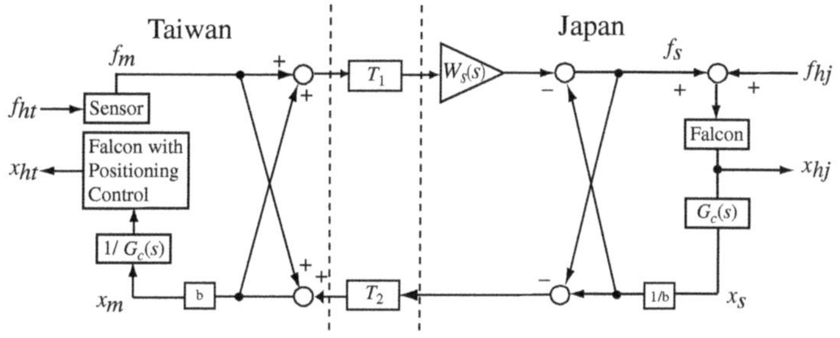

Fig. 2 Control block diagram

slave is consistent with that of the master in a steady state, because we use force-position feedback. Niemeyer's algorithm, in contrast, would drift position, because it was force-velocity feedback. It was not suitable for handshake exhibition. The details of the control algorithm are referred to in [8].

On the Taiwanese side, the detected operating force f_{ht} is transmitted to the Japanese side as force f_s through the scattering matrix and phase compensation filter $W_s(s)$. On the Japanese side, the resultant force of f_s and applied force f_{hj} by a Japanese operator move the Falcon. Accordingly, a Japanese operator can feel his/her force f_{hj}, and the response motion x_{hj} of the Falcon.

The position signal x_s, filtered by the compensation filter $G(s)$, is transmitted to the Taiwanese side as position x_m. After filtering $1/G_c(s)$, commanded position x_r was acquired. Since $G_c(s)$ and $1/G_c(s)$ are complementary, x_{hj} and x_r are almost the same. We implement the position control for the Taiwanese Falcon so that actual position x_{ht} will be consistent with x_r. Thus, the Taiwanese operator can recognize the Japanese Falcon's motion x_{hj}. The program is very simple so that the core routine, which consisted of a scattering matrix, a phase control filter, a phase compensation filter, and positioning proportional control, is only less than 15 lines.

Communication between the two handshake gadgets is performed by P2P socket communication of TCP. Although it runs every 2 ms, traffic is not busy because the data are restored to one packet.

Note that we use the Taiwanese Falcon as the positioning device, although the Falcon is usually used only as a generator of the reaction force. Moreover, the Taiwanese Falcon does not work directly by a Taiwanese operator's force, but by Japanese motion, which is exited by Taiwanese operation force.

3 Demonstration and Results of Questionnaires

3.1 Response of Subject

A Falcon was set up at the Taiwanese and Japanese locations. The gadgets were exhibited at the Niconico Chokaigi April 26–27, 2014. Subjects for the exhibition

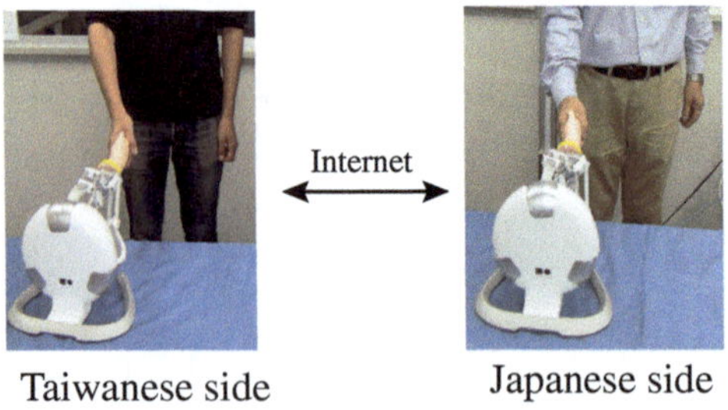

Fig. 3 Image of usage of handshake gadget

included more than 220 persons in both countries. They enjoyed not only handshakes, but also other forms of communication, such as conversation and a 5–10-min quiz, because the purposes of the event were to deepen the friendship between Taiwan and Japan. Some famous individuals and characters also enjoyed the gadget, as may be seen in Fig. 3. The handshake gadget can be seen in the bottom of the photographs.

3.2 Results of Questionnaire

During the exhibition, we asked 219 subjects in both countries to respond to a questionnaire. We included the question: "Was the experiment exciting?" to investigate participants' feelings directly. Answers were given on a 7-level scale from positive to negative. The results are shown in Fig. 4a. As to whether the experiment was "exciting," more than 96 % of subjects answered positively. We could thus see that participants had great interest in our gadget. (The number of answerers is different from the number of participants.)

Between Taiwan and Japan, the communication latency, called round trip time (RTT), was not avoided. Therefore, it was possible that operators would feel the time delay when implementing tele-control via the Internet. The percentage of those who reported feeling the time delay was only 18 % (as shown in Fig. 4b) despite the fact that the RTT was about 60 ms. More than 70 % answered that they had good operability when doing experiments with our system (Fig. 4c). As a result, despite the existing time delay, our proposed system could guarantee good performance.

The primary interest for us in this exhibition was whether or not subjects could feel the force and motion mutually between Taiwan and Japan. The results are shown in Fig. 4d. 106 persons responded that they could feel their partner's force and motion (scale 1, 2, 3), while 42 persons responded negative answer (scale 5, 6, 7). Thus, we concluded that we realized a handshake gadget, as both people mutually felt they were holding the other's arm. The reason why they could not feel partner's force would be weak generative force of Falcon. Falcon's force is less than 10 N so that the human's force overcomes it.

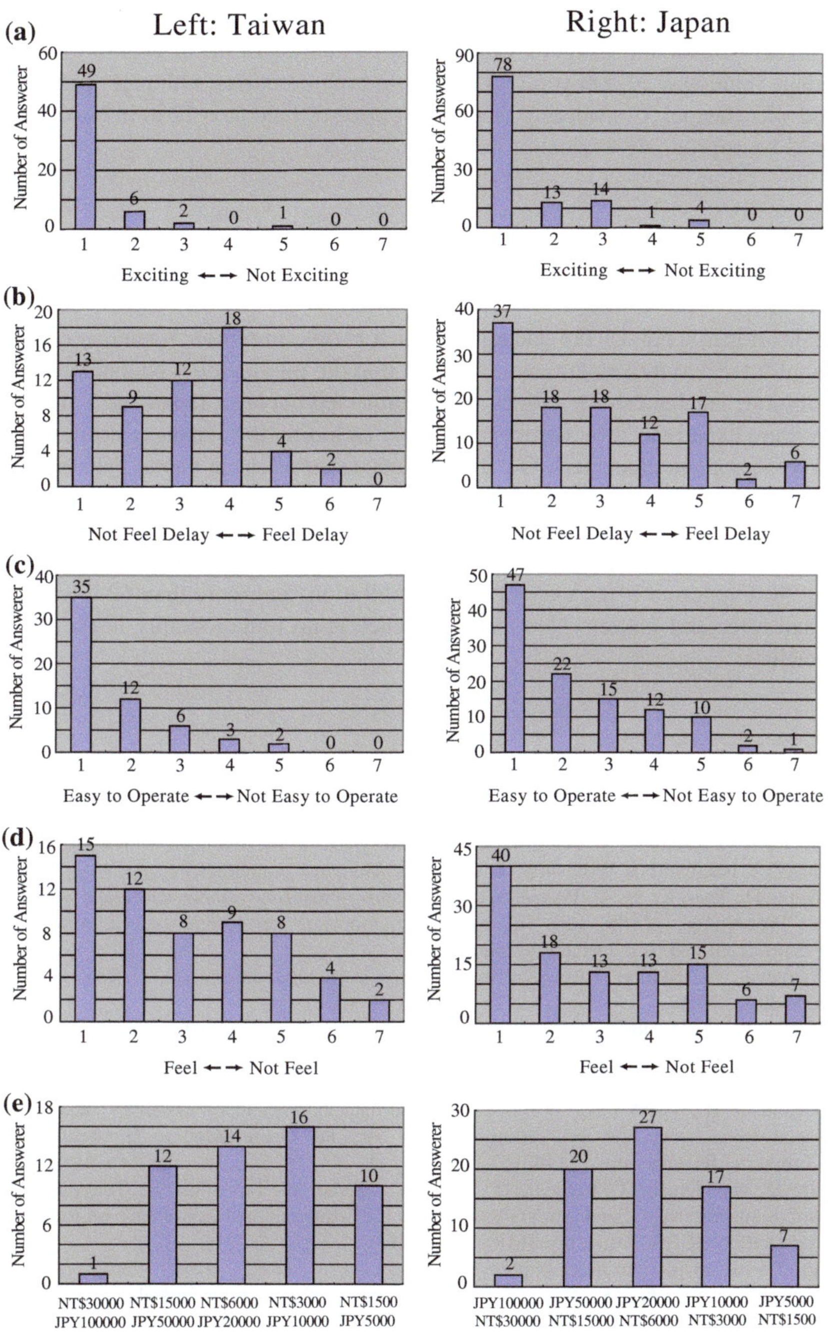

Fig. 4 Results of questionnaire

Finally, we had set the question: "How much can you buy this gadget?" so that we could estimate this gadget's value. In the evaluation of the answer in Fig. 4e, the median value was 20,000 yen (NT$6000) in both countries, although Taiwanese's trend was cheaper. Considering that the list price of Falcon is $250 in Novint HP, it would be appropriate evaluation.

4 Conclusion

We developed a handshake gadget by which people in remote locations could feel mutual force and movement and exhibited it at the international Niconico Chokaigi exhibition. The handshake gadgets were connected between Japan and Taiwan, and they were experienced by number of people including the Japanese Prime Minister and some popular characters. Our gadgets consisted of cheap haptic Falcon devices, force sensors, and our developed software; in other words, anyone could produce these gadgets using simple parts easily available online. At the exhibition, we asked subjects to complete questionnaires about their experience with the system. More than 98 % reported being excited by the exhibition, and more than 62 % could feel mutual force and motion. Now, tele-existence is no longer simply in the realm of specialists.

References

1. Ouchi, K., Hashimoto, S.: Handshake telephone system to communicate with voice and force. In: IEEE International Workshop on Robot and Human Communication, pp. 466–471 (1997)
2. Wang, D., Tuer, K., Ni, L., Porciello, P.: Conducting a real-time remote handshake with haptics. In: Proceedings of the 12th International Symposium on Haptic Interfaces for Virtual Environment and Teleoperator Systems (2004)
3. Niemeyer, G., Slotine, J.J.E.: Telemanipulation with time delays. Int. J. Rob. Res. **23**(9), 873–890 (2004)
4. Kunii, Y., Hashimoto, H.: Tele-handshake using handshake device. In: Proceedings of Conference on Industrial Electronics, Control and Instrumentation (IECON), vol. 1, pp. 179–182 (1995)
5. Nakanishi, H., Tanaka, K., Wada, Y.: Remote handshaking: touch enhances video-mediated social telepresence. In: Proceedings of the SIGCHI Conference on Human Factors in Computing Systems, pp. 2143–2152 (2014)
6. Honda, M., Miyoshi, T., Imamura, T., Mima, K., Okabe, M., Terashima, K.: Tele-manipulation with humanoid robot hand/arm via internet. In: IEEE International Conference on Robotics and Automation (ICRA 2013), pp. 3618–3624 (2013)
7. http://www.novint.com/index.php/novintxio/41
8. Miyoshi, T., Terasima, K., Buss, M.: A design method of wave filter for stabilizing non-passive operating system. In: Proceedings IEEE International Conference on Applications, pp. 1318–1324 (2006)

Haptic Snake: Line-Based Physical Mobile Interaction in 3D Space

Byung-Kil Han, Seung-Chan Kim, Semin Ryu and Dong Soo Kwon

Abstract This paper introduces a shape-rendering method called haptic snake for interacting with virtual objects in air. Haptic snake is composed of serially linked line segments and is controlled using an active contour model known as a snake algorithm. Through an interface controlled using haptic snake, the user can feel a virtual presence in a 3D space from the changes in its shape and the force exerted from the movement of the interface. For the further work, kinesthetic feedback structure will be developed and justified to develop ungrounded mobile haptic interface.

Keywords Haptic interface · Human-computer interaction · Mobile device

1 Introduction

In recently developed handheld devices, haptic feedback is applied for additional information transfer channel and realistic interaction. In these applications, most haptic information is provided to the fingertips in the form of mechanically ungrounded kinesthetic feedback [1–3]. Furthermore, for a more realistic interaction with a geometric 3D object in a 3D space, several researchers have focused on haptic devices to generate kinesthetic feedback to the entire hand [4].

B.-K. Han · S.-C. Kim · S. Ryu · D.S. Kwon (✉)
Korea Advanced Institute of Science and Technology, Daejeon, Republic of Korea
e-mail: kwonds@kaist.ac.kr

B.-K. Han
e-mail: hanbk@robot.kaist.ac.kr

S.-C. Kim
e-mail: kimsc@robot.kaist.ac.kr

S. Ryu
e-mail: ryusm@robot.kaist.ac.kr

H. Kajimoto et al. (eds.), *Haptic Interaction*, Lecture Notes
in Electrical Engineering 277, DOI 10.1007/978-4-431-55690-9_50

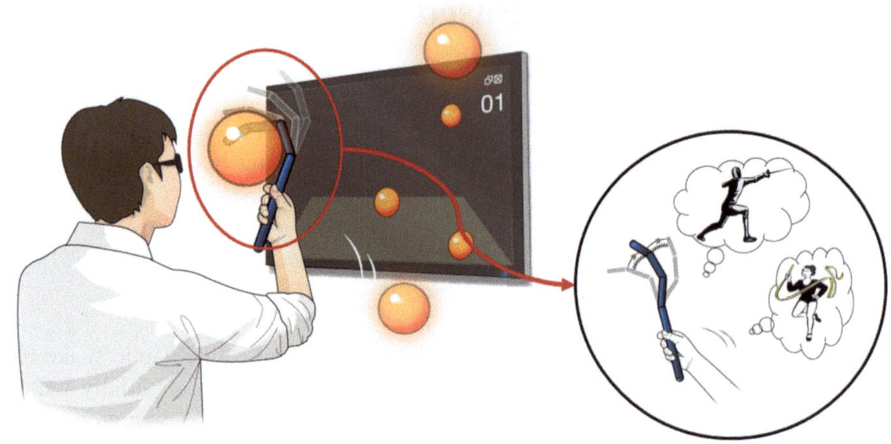

Fig. 1 Conceptual diagram: when the interface makes contact with a virtual object, it deforms to enclose the object

As an extension of these researches, this paper proposes a novel haptic interaction method called haptic snake, which can transmit haptic feedback to the whole hand. A haptic snake is a line-based object interaction method applicable to an articulated mobile interface for interacting with a 3D space. Haptic snake is composed of multiple serially linked line segments, where each segment is controlled using an active contour model [5] for contour segmentation of the applied 2D images. An interface controlled using this method can allow direct interactions with a virtual object by deforming its shape (Fig. 1). In this way, the user can perceive the presence of a virtual object in air through visual feedback from its deformed shape from dynamic movements of the interface.

2 Proposed Method

2.1 Working Principle

The proposed haptic snake was motivated by an active contour model, which is an energy-minimizing spline used to find the contour of an image. For haptic snake, the energy is defined for the distance between a virtual object and each snake element (Fig. 2). From the minimizing process in a time sequence, every snake element moves toward a direction that minimizes the functional energy. Therefore, when the user moves haptic snake to a virtual object, it automatically deforms its shape to enclose the virtual object.

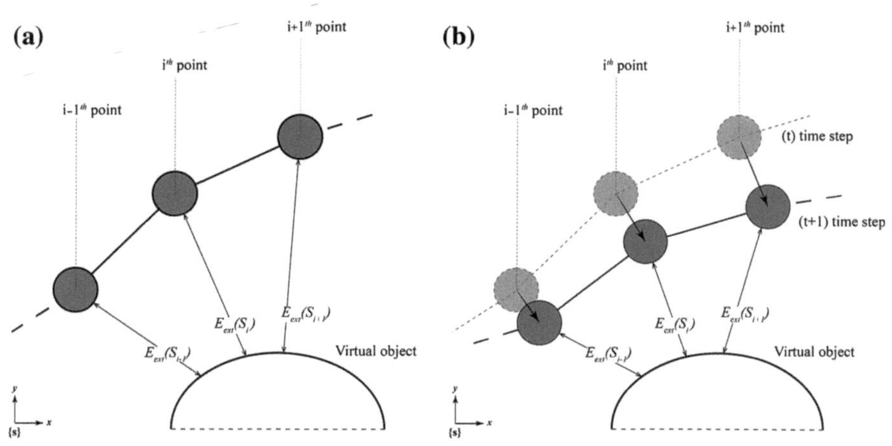

Fig. 2 Structural diagram of external energy (**a**) and its movement (**b**): the snake element moves toward the location that minimizes the energy of each snake element

2.2 *Hardware Implementation*

For physical mobile interaction, the proposed method is implemented in a 3D depth camera and an articulated interface prototype. The interface is a 4R serial manipulator, and each joint was controlled by an AX-12+ motor/encoder system manufactured by Robotis, Co.

3 Application

The proposed method can be applied to a variety of contexts for haptic interactions in a 3D space. First, the external functional energy of haptic snake determines the specific behavior of the snake elements. This energy can be modulated by the user's intention to make haptic snake attract or repel a certain virtual object. Haptic snake can be applied to smart TV applications for selecting a certain content among the numerous contents in 3D space.

4 Conclusion

In this paper, a novel shape-rendering method for an articulated mobile interface, called haptic snake, was proposed. The development of haptic snake was motivated by the active contour model used for the contour segmentation of 2D images. An interface using the haptic snake algorithm changes its shape dynamically with respect to the contour of the virtual object in air. Therefore, the user can perceive the presence of the virtual object based on its deformed shape. The ultimate goal of the present work is ungrounded haptic interface in 3D space. For the further work, force feedback modeling for ungrounded haptic interaction will be developed and verified with perceptual study to achieve this goal. We anticipate that the proposed method can provide an additional means of immersive interactions in a 3D space.

Acknowledgements This work was supported by the ICT R&D program of MSIP/IITP [4-000-11-001, Human Friendly Devices (Skin Patch, Multi-modal Surface) and Device Social Framework Technology].

References

1. Yano, H., Miyamoto, Y., Iwata, H.: Haptic interface for perceiving remote object using a laser range finder. In: EuroHaptics Conference, 2009 and Symposium on Haptic Interfaces for Virtual Environment and Teleoperator Systems. World Haptics 2009. Third Joint, 2009, pp. 196–201. IEEE (2009)
2. Iannacci, F., Turnquist, E., Avrahami, D.: Patel SN the haptic laser: multi-sensation tactile feedback for at-a-distance physical space perception and interaction. In: Proceedings of the SIGCHI Conference on Human Factors in Computing Systems, pp. 2047–2050. ACM (2011)
3. Han, B.-K., Kim, S.-C., Lim, S.-C., Pyo, D., Kwon, D.-S.: Physical mobile interaction with kinesthetic feedback. In: 2012 IEEE Haptics Symposium (HAPTICS), pp. 571–575. IEEE (2012)
4. Swindells, C., Unden, A., Sang, T.: TorqueBAR: an ungrounded haptic feedback device. In: Proceedings of the 5th International Conference on Multimodal Interfaces, pp. 52–59. ACM (2003)
5. Kass, M., Witkin, A., Terzopoulos, D.: Snakes: active contour models. Int. J. Comput. Vision **1** (4), 321–331 (1988)

Panoramic Movie-Rendering Method with Superimposed Computer Graphics for Immersive Walk-Through System

Hikaru Takatori, Hiroaki Yano and Hiroo Iwata

Abstract This paper describes a technique for developing an immersive walk-through system that implements a rendering method for superimposing computer graphics onto a panoramic movie in an immersive walk-through system. The system is composed of a locomotion interface and an immersive spherical display. In this study, we analyze the camera path from the movie to stabilize the orientation of the camera for rendering an immersive image. Further, the system superimposes computer graphics (CG) onto the image according to the system coordinates from the movie. We implemented the proposed rendering method to realize a free viewpoint image with the movie and the CG by moving the viewpoint.

Keywords Locomotion interface · Immersive display · Walk-through system · Evacuation simulator

1 Introduction

An extensive understanding of an evacuee's psychology and behavioral traits is essential in order for emergency workers to decide on an effective evacuation route and to develop effective guidance measures in the event of an emergency. To investigate human characteristics under extreme circumstances, it is pertinent to conduct experiments observing participants in the midst of a reproduced disaster. However, it is difficult to carry this out in the real world because of the great

H. Takatori (✉)
School of Integrative and Global Majors, University of Tsukuba, Tsukuba-shi, Ibaraki, Japan
e-mail: takatori@vrlab.esys.tsukuba.ac.jp

H. Yano · H. Iwata
Faculty of Engineering, Information and Systems, University of Tsukuba, Tsukuba-shi, Ibaraki, Japan
e-mail: yano@iit.tsukuba.ac.jp

H. Iwata
e-mail: iwata@kz.tsukuba.ac.jp

H. Kajimoto et al. (eds.), *Haptic Interaction*, Lecture Notes in Electrical Engineering 277, DOI 10.1007/978-4-431-55690-9_51

dangers involved. Therefore, a simulator can be indispensable for representing a disaster in virtual space to observe realistic behavior during a simulated evacuation.

According to the disaster prevention manual [1] provided by the Fire and Disaster Management Agency, it is advisable to walk, rather than to drive a car or some other vehicle, during an evacuation. Given this, we can say that it is important to reproduce the movement of walking in a virtual evacuation simulator. Hence, the evacuation simulator should have ability to generate haptic sensation of contacting with terrain and user's feet. Also, the ability to generate visual sensation of virtual environment, which includes not only central vision, but also peripheral vision, is required to show evacuation paths. In the initial stage of development of an evacuation simulator, changing the view during a simulated walk becomes very important to the virtual experience so that participants can recognize their surroundings. Thus, generating a quality image that is consistent with the movement from walking is beneficial for the simulator.

In this paper, we describe a technique for rendering a panoramic movie with superimposed computer graphics (CG). This technique is a newly developed method for displaying an image in the evacuation simulator that is composed of a locomotion interface and an immersive display.

2 Problem to Be Solved

In this study, the following objectives will be addressed:

1. Stabilizing the video images.
2. Minimizing the pan rotation of the video caused by changing the direction when recording.
3. Superimposing CG on the video correctly.
4. Generating a panoramic movie that is adapted to the user's walking speed and moving direction.

Stability, the first objective, is a prerequisite to the third, and stable images reduce the motion sickness frequently experienced from using the immersive display. Moreover, when a user changes direction, what should be rotated is the user, not the image. Therefore, the second objective is needed. The third objective is required in order to generate CG composition image without incongruity. Finally, for a realistic walk-through, the fourth objective is indispensable.

3 Prototype System

This section describes the hardware organization for the prototype system developed to address the objectives listed above.

Fig. 1 RearDome and TorusTreadmill

- Camera
  We used Ladybug2, a 360-degree spherical-imaging digital video camera manufactured by PointGrayResearch in order to capture the video of an evacuation route. The system was installed on Patrafour, an electric wheelchair manufactured by KantoAutoWorks. The camera was fixed on an aluminum frame at a height of 1.5 m.
- Immersive display
  RearDome [2] (Fig. 1), developed at our laboratory, was used for the spherical immersive rear projection display. This system is composed of a 1.6-m spherical screen, four projectors, and a PC. This display size is 3.52 m in width, 3.33 m in depth, and 3.10 m in height.
- Locomotion interface
  We conducted the study using TorusTreadmill [3] (Fig. 1), an omnidirectional treadmill. User can walk freely on the treadmill in any direction while staying in the same place. In this study, the user can feel haptic sensation as if he/she walks on a flat surface. The treadmill detects the user's motion and direction, and it revolves and rotates to keep the user at the center of the topside (Fig. 2). The installation area was 2.45 m wide and 2.6 m deep. Thus, this treadmill could be installed under the RearDome. The performance speed is limited to 1.4 m/s.

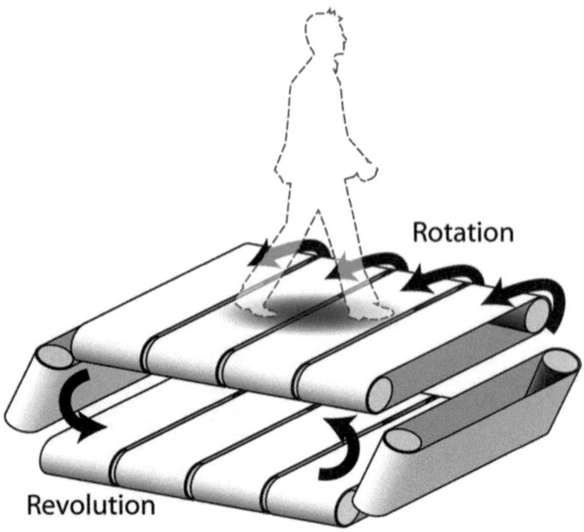

Fig. 2 Operation principle of TorusTreadmill

4 Proposed Method

4.1 Recording the Video of the Route

A manually controlled electric wheelchair traversed the evacuation route taking a 360-degree panoramic video. The resolution of the video was 2480 pixel columns by 1024 pixel rows.

4.2 Camera Path Estimation

We used MatchMover2014, software developed by Autodesk to estimate the camera path by tracking feature points automatically or manually. This technique, called Match Moving, is used widely in the film industry for superimposing CG onto a movie.

Because the video file is a panoramic image of 360°, it cannot be processed with MatchMover2014, which assumes a video input from a monocular camera. Therefore, we coded a program that preprocesses the video. The program we developed draws a sphere with OpenGL and maps the video frame texture onto it, generating an image group captured inside the sphere. The image group is used as the input for MatchMover2014. By processing the images with Matchmover2014, the position and posture of the camera is estimated for every frame, and it is outputted as an XML file.

4.3 *Compositing Process and Projection*

The process described below was implemented in the simulator program. The video file provided a 180-degree vertical angle and a 360-degree horizontal angle. Therefore, users experience the view from the camera on the Patrafour by looking from the center of the sphere that is mapped to the video frame texture. The program stabilizes the video by rotating the sphere in an inverse direction from camera's posture data written in the XML file. The data from the camera path are also used for deciding the viewpoint for drawing the CG. Synchronizing the CG with the video movement is achieved by consistently moving the viewpoint in CG space in tandem with the data from the camera path.

The program generates a composite panoramic image and writes it into the memory for offscreen rendering. The image is mapped on polygons for distortion correction in the spherical screen, and finally, the virtual environment is projected on the screen.

4.4 *Realization of the Walk-Through*

This section provides a comprehensive explanation of the system's operation.

The user's movement is detected with Kinect, an RGB-D sensor manufactured by Microsoft, attached to the TorusTreadmill. When the viewpoint in the virtual environment changes to synchronize with the user's movement, the system cannot generate the corresponding view along the camera path if it merely plays back the video. Therefore, our proposal method chooses the video frame from camera position that is nearest to the user's position in the virtual environment. Moreover, the composite panoramic image is mapped on the inward side of the sphere so that the viewpoint can to some degree move freely inside the sphere.

In summary, the viewpoint in the virtual environment moves in tandem with the user's movement. Next, the nearest video frame is chosen. Then, a textured sphere is drawn at the estimated camera position. As a result, the user is able to move about freely within the sphere (Fig. 3). Figure 4 outlines of the system's operation.

5 Performance Evaluation

5.1 *Frame Rate*

We tested the walk-through system and recorded the frames per second (FPS) for measuring the system's rendering performance. In the experiment, participants walked along the camera path (forward and backward), and against the camera path (right and left), and the FPS was recorded three times in each direction.

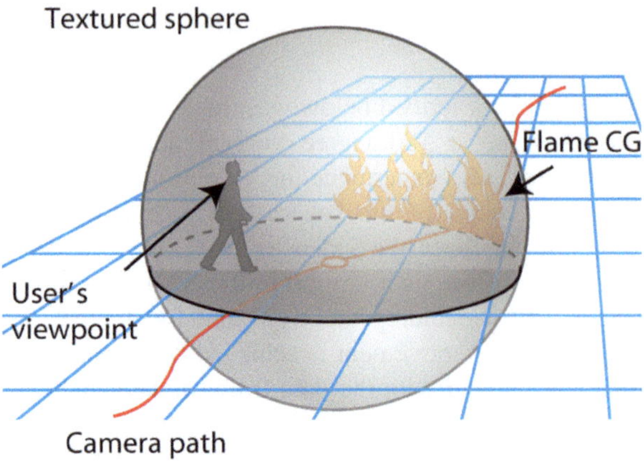

Fig. 3 Relation between the viewpoint, the sphere, and the camera path

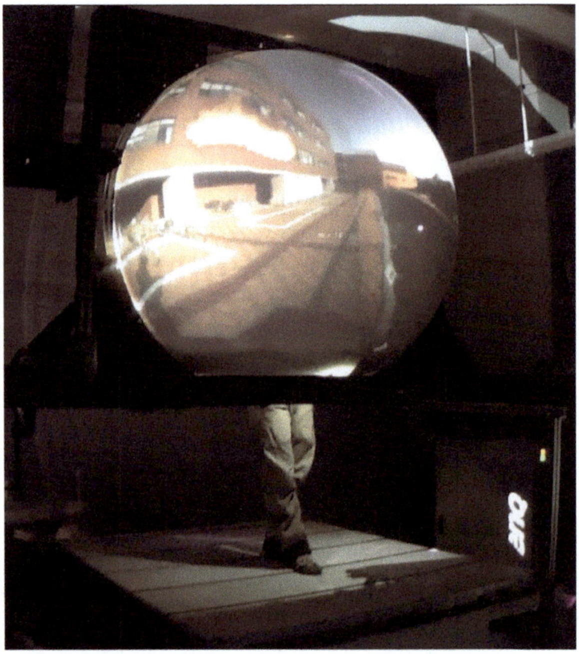

Fig. 4 View from outside of the system

The following graph (Fig. 5) shows the average FPS.

The results for movement against the camera path (left and right) show a high rendering speed relative to the results from walking along the camera path. This suggests that when the program is not loading new video frames, the FPS is high. On the other hand, the results from the backward motion are especially slow. This is because OpenCV—the computer-vision library used to open the video—is not designed to play backward. To resolve this problem, it will be practical for the system to open the video frames as texture in advance.

Fig. 5 Rendering speed of the walk-through system

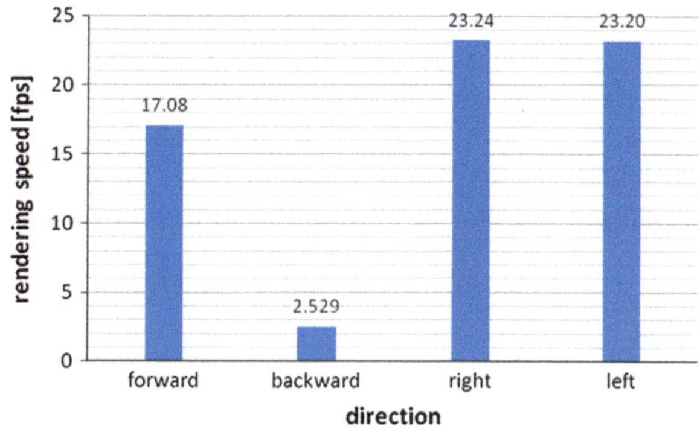

Fig. 6 Walking speed of the walk-through system

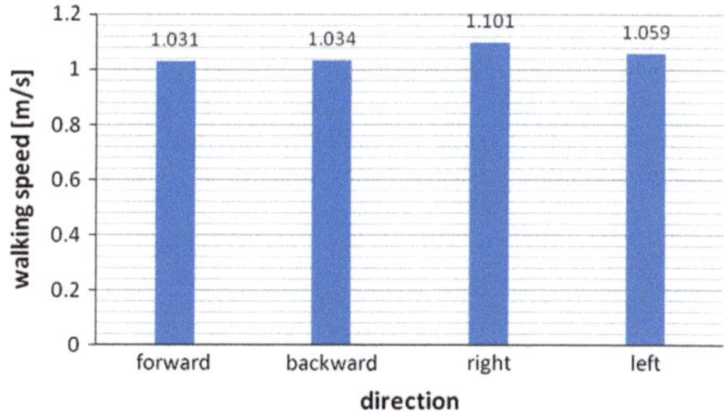

5.2 *Maximum Walking Speed*

We recorded maximum command values generated for TorusTredmill and calculated the highest walking speed. At this experiment, participant walked at top speed, as long as the gate motion is normal, and the command value was recorded 5 times of each direction.

Next graph (Fig. 6) describes the average of the walking speed.

The graph shows that the maximum speed of every direction is over 1.0 m/s, but this is not enough because it is generally said that the normal walking speed is approximately 1.4 m/s. We are trying to improve the user sensing phase now for reaching the limitations in hardware.

6 Conclusion

In this paper, we clarified the issues pertaining to superimposing CG onto panoramic video, and a software solution was described. In particular, this solution offered a way to stabilize the panoramic video and superimpose CG onto the video by estimating the camera path using the Match Moving technique. Furthermore, we implemented a prototype system composed of a locomotion interface and a spherical immersive display, and evaluated its performance. Moreover, we indicated that our system is a new and viable type of virtual evacuation simulator.

Improvements to this research include the development of a more photorealistic simulator that will be used as a platform for behavior analysis.

References

1. The Fire and Disaster Management Agency.: Important information on countermeasures against earthquake disasters, 28 Aug 2007. Available from http://www.fdma.go.jp/bousai_manual/e/. Accessed 10 July 2014
2. Nakagawa, K., Yano, H., Iwata, H.: Development of rear projection full surround spherical display by using 3D short focus projectors (in Japanese). In: Proceedings of the Virtual Reality Society of Japan, Annual Conference, vol. 18, pp. 104–107 (2013)
3. Iwata, H., et al.: Torus treadmill with modular structure. Trans. Virtual Reality Soc. Japan **18**(3), 197–206 (2013). (in Japanese)

Air Tap: The Sensation of Tapping a Rigid Object in Mid-Air

Nobuhisa Miyamoto, Kazuma Aoyama, Masahiro Furukawa, Taro Maeda and Hideyuki Ando

Abstract In developing wearable computing technology, it is often desirable to produce the sensation of tapping a virtual object in mid-air. Such a tapping sensation consists of not only tactile sensations at the fingertips but also force feedbacks to the musculoskeletal system. Therefore, physical interactions with objects ostensibly involve an integration of tactile sensations with force feedback to the musculoskeletal system when perceived objects are contacted. In this study, we propose a method that combines electro-tactile stimulation (ETS) for tactile receptors with functional electro-stimulation (FES) for the musculoskeletal system. The results of an experimental study show that simultaneous perception was reportedly obtained when ETS was delayed by 25 ms after FES when tapping a rigid virtual object.

Keywords Haptic · Tap response · Electrical stimuli · Delay of stimuli

N. Miyamoto (✉) · K. Aoyama · M. Furukawa · T. Maeda · H. Ando
Graduate School of Information Science and Technology, Osaka University, Osaka, Japan
e-mail: miyamoto.nobuhisa@ist.osaka-u.ac.jp

K. Aoyama
e-mail: aoyama.kazuma@ist.osaka-u.ac.jp

M. Furukawa
e-mail: m.furukawa@ist.osaka-u.ac.jp

T. Maeda
e-mail: t_maeda@ist.osaka-u.ac.jp

H. Ando
e-mail: hide@ist.osaka-u.ac.jp

K. Aoyama
Japan Society for Promotion Science Research Fellowship for Young Scientist (DC1), Tokyo, Japan

M. Furukawa · T. Maeda · H. Ando
Center for Information and Neural Networks (Cinet), National Institute of Information and Communication Technology, Osaka, Japan

H. Kajimoto et al. (eds.), *Haptic Interaction*, Lecture Notes
in Electrical Engineering 277, DOI 10.1007/978-4-431-55690-9_52

1 Introduction

To develop wearable computing technology, it is valuable to represent the sensation of tapping virtual objects in mid-air. Sending vibrations is understood to be the easiest method of simulating contact with a virtual object. To do so, users wear only a small-sized vibration actuator to provide vibro-tactile sensations to the fingers when they encounter a virtual object [1]. However, this method cannot reproduce the sensation of contact with an object, because the stimulation is different from real mechanical force feedback, and users must interpret the vibration as the existence of a virtual object. Hence, vibro-tactile sensations are merely symbolic in such cases. Therefore, in addition to tactile sensations at the fingertips, force feedback to the musculoskeletal system is required to accurately represent the sensation of tapping. By integrating stimulation from tactile sensations with force feedback to the musculoskeletal system, users will perceive the sensation of contacting an object. The arm exoskeleton [2] and the wearable master [3] can offer force feedback mechanically in mid-air. However, these structures are complicated, and it is difficult to make them smaller and lighter because small and light actuators cannot generate a large force. Therefore, we focus on electrical stimulation using a lightweight device as a method for stimulating a tactile sensation and providing force feedback. Recent studies are available for a device that provides force feedback using functional electric stimulation (FES) and another that represents tactile sensations with electrical stimulation to the tactile receptors. However, these two methods have not been used together for generating a tapping sensation.

In previous work, Kajimoto et al. [4] proposed electro-tactile stimulation (ETS) as a method for generating tactile sensations. They distributed electrical currents by applying a model to configure the positions of electrical nodes and neural activity to a tactile mechanoreceptor. FES provides electric stimulation to the musculoskeletal system by innervating the muscle with pulse-like electrical currents to stimulate and control the movements of the human body. FES is typically used in the field of medicine to enable the control of body parts that are paralyzed or undergoing rehabilitation. Tamaki et al. [5] presented PossessedHand, another system that uses FES. Their system can control the movements of the joints in a finger by electrical stimulation of the antagonistic muscles for each finger muscles that are found in the forearm. Iwami et al. [6] proposed a teleoperating system with a haptic interface using FES. Their system controls the bending movements of the elbow and wrist joints.

In this paper, we focus on the effects from representing the sensation of tapping using ETS and FES simultaneously. To the best of our knowledge, there is no existing approach that offers such haptic sensations using both ETS and FES. We propose a method for representing the tapping sensation of virtual objects using electrical stimulations for both tactile sensations and the musculoskeletal system. We believe that tactile sensations and force feedback are experienced simultaneously when objects are tapped in the real world. With our proposed method, we are able to produce the sensation of tapping a rigid object in mid-air by having users perceive the two stimulations together. However, FES is prone to minor time latencies [7]. Thus, in this paper, we investigate the time delay in FES for finger movements and determine whether the sequence and the delay of stimuli for tactile sensations and the musculoskeletal system affect the response from tapping rigid virtual objects.

2 Experiment and Method

In order for our proposed method to represent the response from tapping a rigid virtual object, we conducted two experiments that investigated the latency of the perception of force with FES and its effect on the response from tapping a rigid virtual object with time delays in synchronizing FES and ETS.

2.1 Electrical Stimulation and Finger Position Detection

In all of the experiments conducted in this paper, a device that provides a constant electrical current was used to provide FES and ETS. FES controlled the extensor indicis muscle in the right hand. The position of electrodes was determined from the results of a preliminary test: the anode on the tendon and the cathode on the muscle spindle. For ETS, previous work shows that a sense of pressure arises at the cathode [4]. In all experiments, the cathode was set on the distal phalanx of the index finger, and the anode was set on the middle phalanx. The waveform for the electrical stimulation was one pulse with a 25-ms duration. The reason for selecting a duration of 25 ms was that the contact time for tapping actual rigid objects was approximately 25 ms in pretesting. Figure 1 depicts the equipment used for the experiments. The photo-resistor was continuously lit with a laser, and when the tapping finger interrupted the laser, the resistance and voltage in the photo-resistor changed. We used this voltage change as a stimulation trigger.

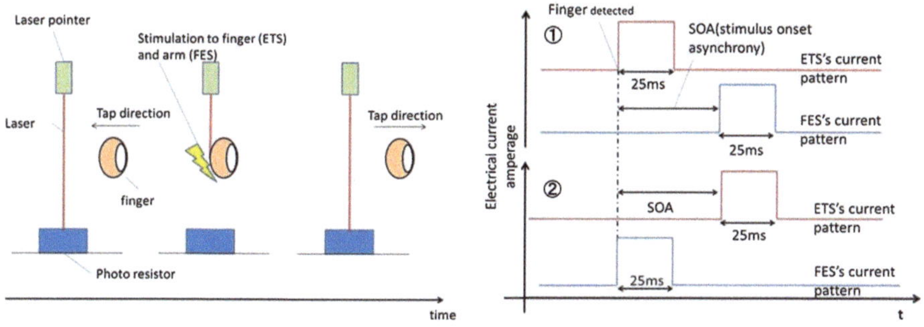

Fig. 1 Measurement equipment (*left*) and stimulation pattern (*right*)

2.2 Experiment 1

In the first experiment, we researched the perception of simultaneity for ETS and FES. This experiment was conducted with three subjects—all 23- to 25-year-old males and all right-hand dominant. The participants rested their right wrist to fix their hand and to restrain any superfluous muscle contractions. Before the experiment, participants adjusted the position of the electro-node such that it could control the extension of the index finger with FES. The stimuli were administered in nine patterns: simultaneous FES–ETS, FES innervating after ETS (12.5-, 25-, 50-, and 100-ms delays) (see Fig. 1), and ETS innervating after FES (12.5-, 25-, 50-, 100-ms delays) (see Fig. 1). Trials were conducted 10 times for each stimulus pattern, totalling 90 trials in random order. Participants wore eye masks and tapped three times during each stimulus before reporting whether the two stimuli were simultaneous. Prior to the experiment, participants practiced a 1-Hz tap using a metronome. Participants answered '1' if they felt the stimuli were simultaneous and '0' if they felt it was not.

2.3 Experiment 2

In the second experiment, we examined the length of the delay between ETS and FES that is most effective for a tapping response. In this paper, we define a rigid body as an object that cannot transform or move. The second experiment was conducted with five participants, using a set-up that was identical to the first experiment. We prepared five stimulation patterns: FES and ETS stimulating simultaneously, FES innervating after ETS (25-ms and 50-ms delays), and ETS innervating after FES (25-ms and 50-ms delays). First, the participants were randomly given three stimuli patterns: simultaneous patterns, and patterns where ETS is stimulated after FES (25-ms and 50-ms delays). The participants were then asked the following three questions: Q1. Did the object feel real?; Q2. Did the object feel correctly shaped?; Q3. Was the object motionless? Further, the participants were asked to rank the three stimuli for each question. Finally, the participants were given a simultaneous pattern and a pattern where FES innervates after ETS (25-ms and 50-ms delays) and likewise asked to rank the stimuli for each question.

3 Results and Discussion

Figure 2 shows the results of the first experiment. The vertical axis represents the average ratio for the perception that the two stimuli were simultaneous, as reported by the participants. The horizontal axis represents the time delay for the ETS after the FES. The negative area on the horizontal axis represents FES innervating after

ETS (that is, the right figure's from Fig. 1). Each dot represents the average for all participants, and the error bar represents the standard error (SE). From this graph, the probability was 80 % that participants reported that the two stimuli were simultaneous with a time delay of 25 ms. This indicates that users will perceive two stimuli as simultaneous if the end of first stimulation coincides with the beginning of the second stimulation. The peak of the graph in Fig. 2 occurs on the positive side because there is a time delay to the perception of force from FES, as we hypothesized.

Figure 3 shows the results of the second experiment. Before the analysis, we converted the ranking into scores such that first place was accorded three points, second place was accorded two points, and third place was accorded one point, and normalized. The vertical axis represents the average score, and the horizontal axis represents the question number. The error bar represents a standard error. The left-side graph in Fig. 3 represents cases where participants most felt the rigid object as a result of the simultaneous pattern. Figure 3 represents cases where participants most felt a tapping sensation corresponding to the rigid object resulting from both simultaneous stimulation and stimulation when ETS was delayed by 25 ms. We believe that one form of stimulation masks the other when they overlap. It is for this reason that a delay of 25 ms was most effective. From the results in Fig. 3, we can conclude that the delay to the stimuli that users experience as rigidity exists between 0 and 25 ms. This span of time matches the time when the two stimuli were perceived as simultaneous in the first experiment. Participants commented that they felt there was a rigid object that did not move and that, in tapping it, their fingers were repelled. We believe that the explanation for this phenomenon is a change in the momentum of tapping the finger caused by a decrease in the finger's velocity. The amplitude of the stimulus did not change, despite a change in the velocity of the finger. Therefore, the finger received more relative momentum, and the participants perceived their fingers as repelling. In other words, we must change

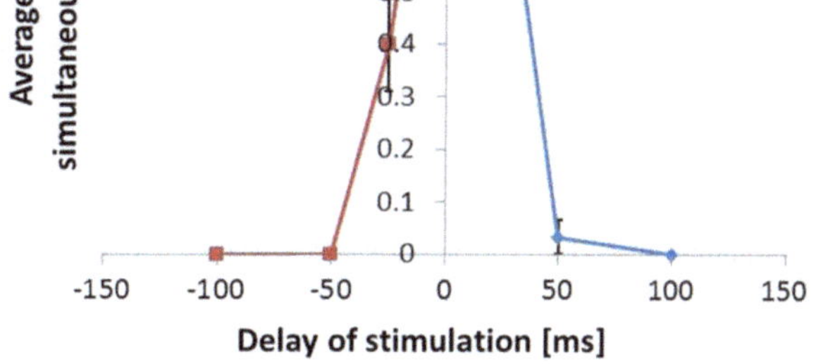

Fig. 2 Results from experiment 1

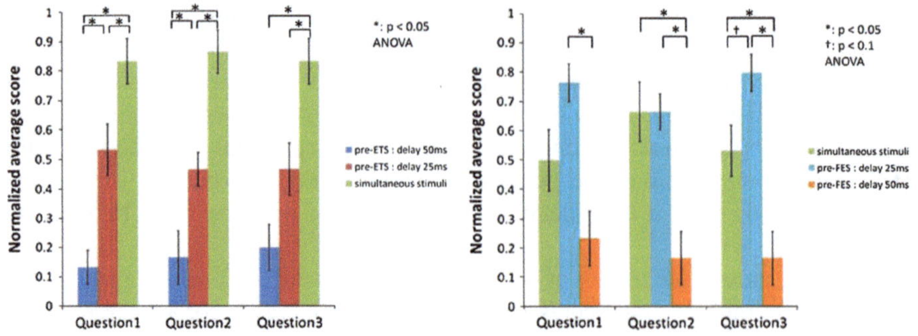

Fig. 3 Results from experiment 2: *left* (pre-ETS patterns: 0, 25, 50 ms), *right* (pre-FES patterns: 0, 25, 50 ms)

the amplitude of the stimulus so that it accords with changes in the velocity of the tapping finger. Participants also commented that they felt that they were tapping something resembling a thin sheet of ice and that they had shattered it when the time delay between the FES and ETS was large. We may thus be able to modify various kinds of virtual objects by changing the time delay. Nevertheless, we demonstrated the need for a time delay so that a virtual object is perceived as a rigid body and where tapping it is perceived as a simultaneous stimulation of the tactile sense and musculoskeletal system.

4 Conclusion

In this paper, we proposed a method for generating the sensation of tapping an object by stimulating the tactile receptor and musculoskeletal system using electrical stimulation. We hypothesized that we could produce the sensation of tapping a rigid object in mid-air by having users perceive the two stimulations as simultaneous. We researched the perception of simultaneity for the two stimulations and the most effective time delay between ETS and FES for the tapping response. The results of the experimental study showed that a pre-FES 25-ms delay was most effective in terms of a tapping response for perceiving the two stimuli—that is, stimulations to the tactile sense and the musculoskeletal system—as simultaneous. Therefore, this delay is necessary for the perception of stimulation by FES when we design methods that simulate tapping rigid objects. In future, we shall attempt to represent more realistic tap responses with an ETS that is better matched to the mechanical movement of a finger stimulated by FES.

Acknowledgments This work is supported in part by the Strategic Information and Communications R&D Promotion Programme (SCOPE) of the Ministry of Internal Affairs and Communications of Japan.

References

1. Immersion Corp., CyberTouch
2. Sledd, A., O'Malley, M.K.: Performance enhancement of a haptic arm exoskeleton. In: 2006 14th Symposium on Haptic Interfaces for Virtual Environment and Teleoperator Systems, pp. 375–381. IEEE (2006)
3. Iwata, H., Nakagawa, H.: Wearable force feedback joystick. In: The Institute of Image Information and Television Engineers, pp. 15–18 (1998)
4. Kajimoto, H., Kawakami, N., Maeda, T.: Electrocutaneous display with receptor selective stimulations. Ins. Electron. Commun. Eng. D **84**(1), 120–128 (2001)
5. Tamaki, E., Miyaki, T., Rekimoto, J.: PossessedHand: techniques for controlling human hands using electrical muscles stimuli. In: Proceedings of the SIGCHI Conference on Human Factors in Computing Systems, pp. 543–552. ACM (2011)
6. Iwami, T., Miura, H., Hasegawa, K., Nakayama, A., Obinata, G., Miyawaki, K., Yanagihara, Y.: A new bilateral teleoperator with force reflection using functional electrical. Robot. Soc. Japan **20**(8), 844–851 (2002)
7. Granat, M.H., et al.: Improving limb flexion in FES gait using the flexion withdrawal response for the spinal cord injured person. J. Biomed. Eng. **15**(1), 51–56 (1993)

Haptic-Enabled English Education System

Minh Phuong Hoang, Jaebong Lee, Hojin Lee, Kyusong Lee, Gary Geunbae Lee and Seungmoon Choi

Abstract For computer-based education, virtual environments equipped with haptic devices have been used to aid the learning process. The main benefit of haptics technology is that it can improve the realism of virtual environments. This paper describes a haptic-enabled English education system for elementary students. We demonstrate two game-based scenarios which allow users to learn English and scientific laws by exploring in virtual environments with a haptic interface.

Keywords Virtual reality simulation · Game-enabled language education · Force feedback devices · Physics-based games · Elementary students

M.P. Hoang (✉) · J. Lee · H. Lee · K. Lee · G.G. Lee · S. Choi
Department of Computer Science and Engineering, Pohang University of Science and Technology (POSTECH), Pohang, Korea
e-mail: phuonghm@postech.ac.kr
URL: http://hvr.postech.ac.kr/

J. Lee
e-mail: novaever@postech.ac.kr

H. Lee
e-mail: hojini33@postech.ac.kr

K. Lee
e-mail: kyusonglee@postech.ac.kr

G.G. Lee
e-mail: gblee@postech.ac.kr

S. Choi
e-mail: choism@postech.ac.kr

H. Kajimoto et al. (eds.), *Haptic Interaction*, Lecture Notes in Electrical Engineering 277, DOI 10.1007/978-4-431-55690-9_53

1 Introduction

English has been a contemporary *lingua franca*, and the consequent use of English as a global language has brought increased interests in English education. In many countries, governments have invested huge expenses to provide more effective English education [7]. In Asian countries, students commonly experience difficulties in learning English due to the linguistic discrepancies in principles and structures between English and their own language [1]. Encouraging students to be active in the learning process is one of the most important requirements in English education.

Many researchers developed computer-aided systems for English education, and some of them use dialog systems [5, 6, 8]. These systems motivate students to be more interactive by recognizing utterances and providing educational feedback with voice in specific mission-oriented scenarios. Haptic feedback can provide more immersive and interactive environments to students by directly delivering innate tactual feedback to a user's body. Several researchers developed haptic education systems by augmenting or replacing the visual and aural information with haptic information and showed potential in providing more intuitive education especially for science [2, 3, 9].

Inspired by previous studies, we developed a novel English education system by integrating a spoken dialogue system and a force-feedback haptic interface to provide an active, intuitive, and immersive learning environment (Fig. 1). Our system allows students to study English in two science-related scenarios, a gravity game and a texture exploration. Students can improve English skills and understand about scientific phenomena in our haptic-enabled English education system.

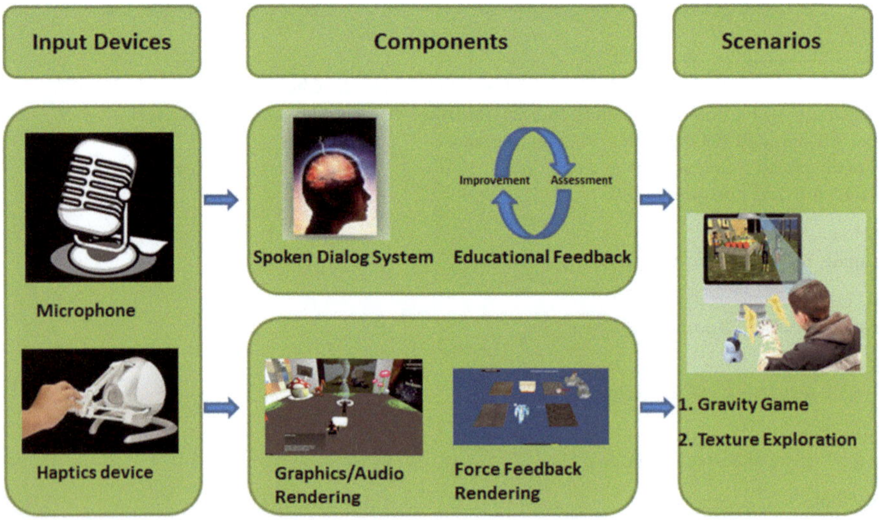

Fig. 1 Overview of our haptic-enabled English education system

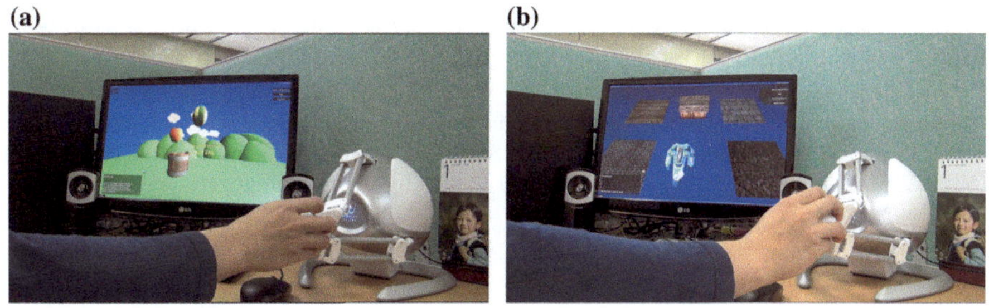

Fig. 2 Demo scenarios using a Novint Falcon. **a** Gravity game. **b** Texture exploration

2 Overview of Demonstration

2.1 Gravity Game

In the gravity game, a student moves a virtual bucket to catch falling objects, such as an apple, a banana, and a watermelon (Fig. 2a). The student can perceive a gravitational force $F = m \cdot g$, where g is the acceleration of gravity, and m is the mass of an object. After the game, our system asks questions. For example, if the system asks "What is the heaviest object?", then the student can answer "A watermelon is the heaviest object." Therefore, students can learn both English conversation and the law of gravity.

2.2 Texture Exploration

In the texture exploration, a student can intuitively learn adjectives while feeling textures of various materials, such as metal, wood, and rock, using a force-feedback haptic interface (Fig. 2b). Surface details of virtual objects are rendered by image-based haptic texturing which is developed by Ho et al. [4]. When the student explores virtual objects using the haptic interface, our system explains verbally the characteristics of the material. The students can also express perceived tactile sensation using appropriate adjectives, for example, "Metal is hard and smooth" and "Rock is rough and uneven."

3 Conclusion

Our haptic-enabled English education system was developed to contribute to better learning of English by providing more intuitive and immersive environments. In the demonstration, we plan to collect comments about the system and accommodate them to upgrade English learning scenarios. In the future, we plan to conduct a formal experiment to show the effectiveness of our system.

Acknowledgement This work was supported in part by a BRL program (2010-0019523) and a Mid-Career Researcher Program (2013R1A2A2A01016907) from NRF and by an ITRC program (NIPA-2010-C1090-1011-0008) from NIPA, all funded by the Korean government.

References

1. Cho, B.: Issues concerning Korean learners of English: English education in Korea and some common difficulties of Korean students. East Asian Learner **1**(2), 31–36 (2004)
2. Grow, D.I., Verner, L.N., Okamura, A.M.: Educational haptics. In: AAAI Spring Symposium: Semantic Scientific Knowledge Integration, pp. 53–58 (2007)
3. Hamza-Lup, F.G., Baird, W.H.: Feel the static and kinetic friction. In: Lecture Notes on Computer Science (Eurohaptics 2012, Part I), vol. 7282, pp. 181–192 (2012)
4. Ho, C., Basdogan, C., Srinivasan, M.: Efficient point-based rendering techniques for haptic display of virtual objects. Presence Teleoperators Virtual Environ. **8**(5), 477–491 (1999)
5. Johnson, W.L., Wang, N., Wu, S.: Experience with serious games for learning foreign languages and cultures. In: SimTecT Conference (2007)
6. Morton, H., Jack, M.A.: Scenario-based spoken interaction with virtual agents. Comput. Assist Lang. Learn. **18**(3), 171–191 (2005)
7. Nunan, D.: The impact of English as a global language on educational policies and practices in the Asia-Pacific region. TESOL Q. **37**(4), 589–613 (2003)
8. Reis, R., Escudeiro, P.: Educational software to enhance English language teaching in primary school. In: International Conference on Information Technology Based Higher Education and Training (ITHET), pp. 1–4 (2011)
9. Sato, M., Liu, X., Murayama, J., Akahane, K., Isshiki, M.: A haptic virtual environment for molecular chemistry education. In: Lecture Notes on Computer Science (Transactions on Edutainment I), vol. 5080, pp. 28–39 (2008)

Visual Vibrations to Simulate Taps on Different Materials

Taku Hachisu, Gabriel Cirio, Maud Marchal, Anatole Lécuyer and Hiroyuki Kajimoto

Abstract This paper presents a haptic visualization technique for conveying material type through visual feedback, expressed as visible decaying sinusoidal vibration resulting from tapping an object. The technique employs cartoon-inspired visual effects and modulates the scale of the vibration to comply with visual perception. The results of a user study show that participants could successfully perceive three types of material (rubber, wood, and aluminum) using our novel visual effect.

Keywords Cursor · Visual simulation of haptics · Vibration · Material

1 Introduction

Using haptic visualization (HV) techniques is more practical to convey haptic events in virtual environments than using complex mechanical devices in terms of cost. However, HV is limited in terms of frequency bandwidth because of limitations imposed by the visual display and perception. Whereas vibration around 500 Hz can

---

T. Hachisu (✉) · H. Kajimoto
The University of Electro-Communications, 1-5-1 Chofugaoka Chofu, Tokyo 182-8585, Japan
e-mail: hachisu@kaji-lab.jp

H. Kajimoto
e-mail: kajimoto@kaji-lab.jp

G. Cirio · M. Marchal · A. Lécuyer
INRIA Rennes Campus de Beaulieu, 35042 Rennes Cedex, France
e-mail: gabriel.cirio@inria.fr

M. Marchal
e-mail: maud.marchal@inria.fr

A. Lécuyer
e-mail: anatole.lecuyer@inria.fr

H. Kajimoto et al. (eds.), *Haptic Interaction*, Lecture Notes in Electrical Engineering 277, DOI 10.1007/978-4-431-55690-9_54

be haptically perceived [1], the maximum frequency perceivable for vision is less than 50 Hz [2].

In this paper, we present a novel HV technique to provide higher frequency of haptic cues. We focus on the vibration resulting from tapping an object, which is an initial haptic cue for grasping the properties of the object [3, 4]. We describe our HV method, which employs a decaying sinusoid model and a cartoonified rendering method, that modulates the model to fit the display and visual perception. The results of a user study show that participants could successfully identify three types of materials (rubber, wood, and aluminum) assigned to virtual objects.

2 Related Work

Pseudo-haptics is a type of visuo-haptic illusion that is especially accompanied by motion [5]. For instance, when moving a PC mouse on a smooth surface, if the velocity of the cursor on the monitor slows down in comparison with that of the mouse, the user feels as if the mouse has become heavier. Interestingly, pseudo-haptics induces actual kinesthetic sensations [6], while other HVs do not always induce haptic sensations. Besides pseudo-haptics, various HV techniques have been developed by modulating the trajectory [7], movement, and size [8] of a visual representations which provide haptic cues.

With the vibrations resulting from contact, we can both identify the material and estimate the stiffness of the object [3, 4]. However, to the best of the authors' knowledge, no attempt has been made to simulate the vibrations in a HV context. As described in the previous section, it is difficult to visually simulate the vibrations because of the limitations imposed by the visual display and perception.

On the other hand, representing the vibrations with a conventional force feedback display was also challenging since it requires high frequency (vibration) as well as low frequency (force). To cope with this difficulty, Wellman and Howe [3] mounted a vibrator on the haptic display. Then, Okamura et al. [4] provided the both by modifying the vibratory parameters to fit the bandwidth of the haptic display. The both studies simulated the vibrations resulting from tapping, by employing the following decaying sinusoid:

$$Q(t) = A(v)\exp(-\mathrm{Bt})\sin(2\pi \mathrm{ft}) \tag{1}$$

Vibration Q is determined by the amplitude A as a function of the impact velocity v, the decay rate of sinusoid B, and the frequency f. These parameters are dependent on the type of material. Figure 1 illustrates the frequency of vibromechanical bandwidth, represented vibrations ([3, 4] and our approach), and the haptic and visual bandwidths.

Okamura et al. [4] demonstrated that it was possible to simulate the material with a vibration model that was designed based on haptic perception and display, which is different from the actual physical phenomenon. Hashimoto et al. [9] implemented

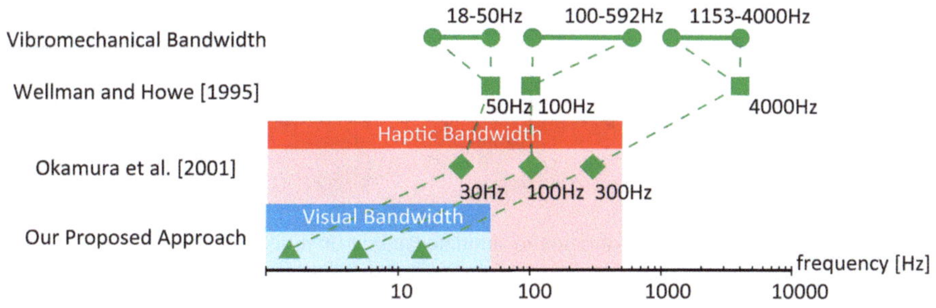

Fig. 1 Frequency of vibration resulting from tapping (i.e., vibromechanical bandwidth), frequency of presenting the vibration, and haptic and visual bandwidths for vibration

a similar concept for a haptically achieved "slow-motion" effect. While these studies focus on scaling along the temporal axis, attempts have also been made to exaggerate haptic reactions, in a process called cartoon-inspired haptic rendering [10]. These authors modulated the vibrations of the impact, which, although not same as reality, appeared to enhance the interaction experience.

3 Visually Simulating Material Vibration

This section describes an implementation of visual vibrations of collision on a two-dimensional monitor. Since the frequency, duration, and amplitude of the real vibrations cannot be visually perceived by users or be displayed on the monitor, we present the following HV method inspired by Wellman and Howe [3], Hashimoto et al. [9], Gleeson and Johnson [10]. The frequency range problem is resolved by scaling down f and B to a visually perceptible range. The amplitude range problem is resolved by amplifying the amplitude.

First, the algorithm defines the position of a virtual object and then draws it on the monitor. The cursor is also drawn according to the user's manipulation using the mouse. When the edge of the cursor touches the surface of the object, the algorithm detects a collision. Upon detection of the collision, the algorithm obtains the velocity of the cursor. Next, the algorithm begins vibrating the cursor according to Eq. 1. We implemented two vibrating modes: (1) B-in mode sets the initial collision position as the center of vibration, as shown in the left panels of Fig. 2. Thus, the cursor penetrates the object. (2) B-on mode vibrates the cursor but does not penetrate the surface of the virtual object, as shown in the right panels of Fig. 2. Thus, the frequency of the cursor appears to be doubled. We expect that the two modes will enable users to understand the elasticity of the contacting object: A hard object is hard to deform, while a soft object easily deforms so that the position of the cursor moves forward. In summary, our algorithm expresses the material properties using four parameters, while there are other possibilities, such as vibrating the object.

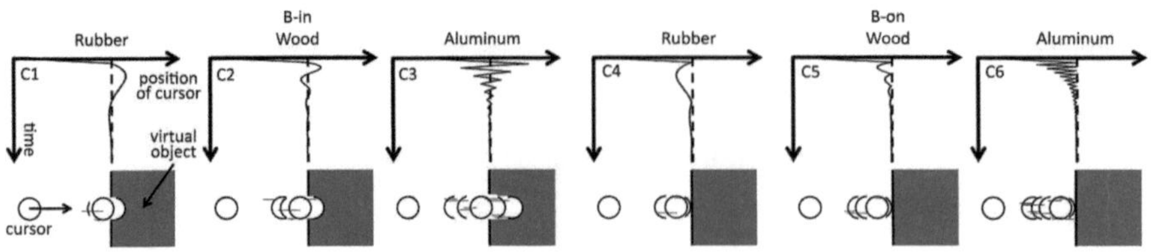

Fig. 2 Illustration of the visual vibration to simulate tapping on different materials (rubber, wood, and aluminum) in two different conditions: B-in (*left*) and B-on condition (*right*)

Table 1 The six sets of parameters and correct answer rate in the experiment

No.	Material	A (s)	B (s^{-1})	f (Hz)	Vibrating	Correct answer rate (%)
C1	Rubber	0.015	6.0	1.5	B-in	79.2
C2	Wood	0.009	8.0	5.0	B-in	61.5
C3	Aluminum	0.020	9.0	15.0	B-in	68.8
C4	Rubber	0.015	6.0	1.5	B-on	40.6
C5	Wood	0.009	8.0	5.0	B-on	53.1
C6	Aluminum	0.020	9.0	15.0	B-on	69.8

We selected three sets of material parameters corresponding to rubber, wood, and aluminum, taken from Okamura et al. [4]. We modified the parameters to enable all the participants to see the vibration, as listed in Table 1. These parameters were determined by haptic experts (the authors) through a preliminary test that, in practice, led to the linear down-scaling of Okamura et al. [4]. Concerning A and B, we knew that the users could notice a visual difference between the three materials. Concerning F, we set the highest frequency as 15 Hz, which is the fastest vibration that the LCD can support and that the users were able to perceive stably. For each material parameter, we prepared two vibrating conditions (B-in and B-on). Thus, we had six conditions.

4 Experiment

4.1 Setup and Procedure

The goal of the experiment was to assess whether participants could interpret our method as intended. We used a 20-inch liquid crystal display (LCD) with resolution 800 × 600, refresh rate 60 Hz, and a three-button infrared mouse. The green mouse cursor and the white virtual object were displayed as shown in Fig. 3. The cursor could only be moved horizontally by sliding the mouse horizontally. We used the six sets of parameters (Table 1). Twelve participants (10 males and 2 females, 22–29 years) were recruited. Participants were divided into two groups (G1 and G2).

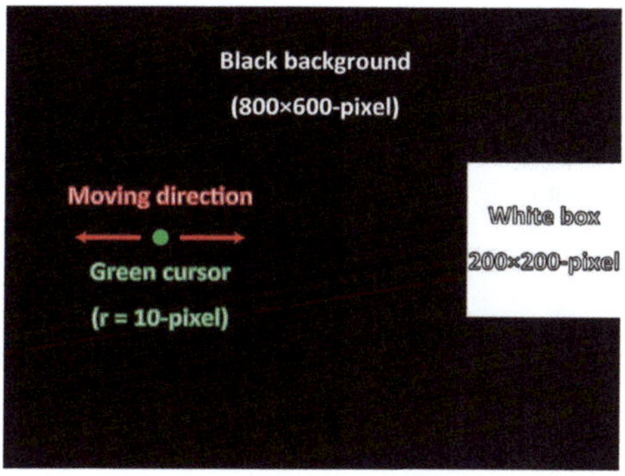

Fig. 3 Screen shown to participants during the experiment

The experiment comprised two successive tests (material identification and realism rating). Because participants were given only three alternatives in the identification test, it is possible that they could have guessed the correct answer without actually feeling the realism. Thus, the objective of the realism rating test was to quantitatively evaluate the realism of the vibratory parameters. After the identification test, the same participants conducted the realism rating test.

In the identification test, participants sat in front of the LCD on the table with their dominant hand on the mouse. Before each trial, participants were asked to set the mouse on the start line marked on the table. The cursor was placed 400 pixels away from the white box. To begin a trial, participants pressed the space bar on the keyboard. During the trials, participants were asked to tap the box with the cursor and identify the material. Participants were allowed to tap the box as many as they wanted within 5 s for each trial. After each trial, the names of the three materials (rubber, wood, and aluminum) appeared on the screen to remind participants of the options, while the cursor and box disappeared. The participants were then asked to use the keyboard to select the material of the box. G1 first performed the trials in the B-in condition, and then in the B-on condition. Before the trials, participants first performed six training trials. They then carried out 24 trials, followed by the same procedure in the B-on condition. Thus, each participant performed a total of 60 trials. G2 performed the trials in reverse order. For each trial, the material of the box was randomly set.

In the realism rating test, the six condition numbers and evaluation scores were displayed at the top of the screen, and the current vibratory condition was marked. Participants were asked to imagine that the box was made of rubber and to tap the box with the cursor. They were then asked to rate the realism of the six conditions on a seven-grade scale (1: definitely not rubber, 7: really made of rubber). During the evaluation, participants were allowed to switch to the other conditions, thereby allowing them to evaluate one condition against another. There was no time limit, and they could tap the box as many as they wanted. After evaluating all six conditions for rubber, participants were asked to imagine that the box was made of wood and then aluminum while they repeated the same procedure. Participants were unaware of the correct answers until end of the experiment.

Table 2 Result of realism rating test: the medians (inter-quartile ranges) of the scores for each condition (shaded cells represent the correct material for each condition)

	C1	C2	C3	C4	C5	C6	Steel-Dwass Test ($p < 0.05$)
Rubber	4 (3.25)	4 (4.25)	1 (0)	3 (2)	2 (1.5)	1 (0)	C1, C2 > C3, C6; C4 > C3
Wood	1 (0)	1 (2.25)	1 (1.25)	6 (1)	5 (2)	1 (1.25)	C4 > C1, C2, C3, C6; C5 > C1, C3, C5
Aluminum	1 (0)	1 (1)	4 (3.25)	1.5 (2.25)	3 (1)	6 (1)	C3 > C1; C5 > C1, C2; C6 > C1, C2, C4

4.2 Results and Discussion

The correct answer rates for all six conditions are given in Table 1. The chance level of the correct answers was 33.3 %. We performed a chi-squared goodness-of-fit test of the correct and found that participants produced significantly more correct answers, i.e., overall percent correct = 62.2 % > 33.3 % ($\chi^2(1) = 215.3$, $p < 0.001$), which means that the visual vibration enabled participants to identify the material of the box.

Table 2 shows the medians and inter-quartile ranges of the scores for each condition and result of Steel-Dwass test. The result indicates that the B-on condition increased the score for the realism of wood (C4 and C5) and aluminum (C5 and C6). Similarly, the B-in condition increased the score for the realism of rubber (C1 and C2).

According to questionnaire after the experiment, a realistic perception was not reported by five participants. By carefully reading the questionnaires, we could categorize them into two groups: (1) not feeling realistic at all and (2) not feeling realistic, but understandable. The former means participants did not understand what the visual vibration was intended to provide. Meanwhile, the latter means they did not feel that the object was realistic, but something that conveyed the essence of reality. In other word, our method was not realistic, but conceivable. Thus, the interpretation of the visual vibration relies on participants' subjective impressions. For the application of our method, we assume that users correctly understand what the designer means by visual vibration after an introduction comprising a brief explanation of haptics or the enrichment of context. In particular, designing context is important because the participants' impression is easily affected by their interpretation of the situation, just as the same pseudo-haptic feedback can be interpreted as friction or weight [5].

5 Conclusion

This paper presented a HV technique, which employs modified decaying sinusoidal vibrations resulting from tapping an object to comply with visual perception. The results of a user study showed that participants could successfully discern three types of materials (rubber, wood, and aluminum) assigned to virtual objects. However, the implementation can be improved by refining parameters, vibrating objects (not cursor), and using a higher refresh rate display such as cathode-ray tube display, which expands the expression of our method.

We applied this technique to a music video game and simulated virtual chromatic percussion instruments (xylophone and glockenspiel) using visual and tactile vibrations [11]. Users could feel the material of different instruments haptically as well as aurally. The King-Kong Effect [12] is another application, in which visual and tactile vibrations are applied to walking in virtual environments. The visual vibration is presented by vibrating a virtual camera that provided a first person view to a user in front of a monitor and improved the sensation of walking. It means that our method can be applied to convey the surface properties of the ground.

References

1. Jones, L.A., Lederman, S.: Human Hand Function. Oxford University Press, Oxford (2006)
2. Hylkema, B.S.: Examination of the visual field by determining the fusion frequency. Acta Ophthalmol. **20**, 181–193 (1942)
3. Wellman, P., Howe, R.D.: Towards realistic vibrotactile display in virtual environments. ASME Dyn. Syst. Control Div. **57**, 713–718 (1995)
4. Okamura, A.M., Cutkosky, M., Dennerlein, J.: Reality based models for vibration feedback in virtual environments. IEEE/ASME Trans. Mechatron. **6**, 245–252 (2001)
5. Lécuyer, A.: Simulating haptic feedback using vision: a survey of research and applications of pseudo-haptic feedback. Presence **18**, 39–53 (2009)
6. Tanikawa, T., Hirose, M.: A study of multi-modal system with visual feedback. In: International Symposium on Universal Communication, pp. 285—292 (2008)
7. Ban, Y., Kajinami T., Narumi, T., Tanikawa, T., Hirose, M.: Modifying an identified curved surface shape using pseudo-haptic effect. In: IEEE Haptics Symposium, pp. 211—216 (2012)
8. Watanabe, K.: VisualHaptics (2002). http://www.persistent.org/VisualHapticsWeb.html
9. Hashimoto, Y., Nakata, S., Kajimoto, H.: Novel tactile display for emotional tactile experience. In: International Conference on Advances in Computer Entertainment Technology, pp. 124—131 (2009)
10. Gleeson, B., Johnson, D.E.: Expressive haptic rendering with cartoon-inspired effects. In: IEEE Haptics Symposium, pp. 191–194 (2010)
11. Hachisu, T., Cirio, G., Marchal, M., Lécuyer, A., Kajimoto, H.: Virtual chromatic percussions simulated by pseudo-haptic and vibrotactile feedback. In: International Conference on Advances in Computer Entertainment Technology, p. 20 (2011)
12. Terziman, L., Marchal, M., Multon, F., Arnaldi, B., Lécuyer, A.: The king-kong effects: improving sensation of walking in VR with visual and tactile vibrations at each step. In: IEEE Symposium on 3D User Interfaces, pp. 19—26 (2012)

Haptic Interface for Shape and Texture Recognition of Remote Objects by Using a Laser Range Finder

Yoshiyuki Yamashita, Hiroaki Yano and Hiroo Iwata

Abstract This paper proposes a system that enables users to recognize the shapes and textures of remote objects with 1-degree of freedom (DOF) haptic feedback. The system consists of a camera platform and a handheld haptic interface. The platform is equipped with a laser range finder, which can be tilted and panned. The handheld interface can generate a 1-DOF reaction force proportional to the distance measured by using a laser range finder (LRF) on the camera platform. The LRF rotates synchronously according to the translational and rotational motion of the handheld interface. In addition, a stereoscopic display indicates a real-time stereoscopic video image of a remote site captured by a stereo camera.

Keywords Virtual reality · Haptic · Tele-existence · Master–Slave system

1 Introduction

The technology for generating the feeling of existing in a remote place is called tele-existence. Tele-existence is realized using a master–slave system with a remote humanoid robot and presenting the information of a remote location, which is consistent with the operation of the user and the remote robot. The user can obtain various sensations of the remote robot through the interface devices. A potential field of application of tele-existence is telemedicine. Further, it is thought that tele-existence can be used in various applications for the general public, such as virtual travel. However, in the existing system, e.g., TERESAR-V, haptic interfaces for

Y. Yamashita (✉) · H. Yano · H. Iwata
Systems and Information Engineering, University of Tsukuba, Tsukuba, Ibaraki, Japan
e-mail: yamasita@vrlab.esys.tsukuba.ac.jp

H. Yano
e-mail: yano@iit.tsukuba.ac.jp

H. Iwata
e-mail: iwata@kz.tsukuba.ac.jp

H. Kajimoto et al. (eds.), *Haptic Interaction*, Lecture Notes
in Electrical Engineering 277, DOI 10.1007/978-4-431-55690-9_55

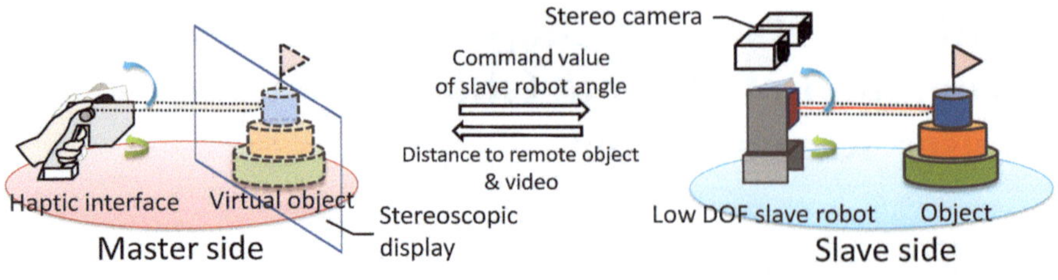

Fig. 1 System concept

tele-existence systems tend to be complicated and expensive [1]. Therefore, we aim to reproduce, to the best possible extent, the sense of touch of a remote object by using a simple mechanism in this study (Fig. 1).

2 System Concept

In order to obtain a sense of touch, a person moves his/her finger toward any object that he/she wants to touch and perceives the reaction force. To realize the sense of touching objects at a remote site while the user remains at the local site, an apparatus for obtaining the information of the shapes and textures of the remote objects and the interface for displaying the obtained information to the user's finger as the reaction force are necessary. A typical high-end tele-existence system consists of a multi-joint manipulator equipped with six axial force sensors and a 6-degree of freedom (DOF) force, and a tactile information display device that is attached to each finger of the user. In contrast, as a low-DOF haptic interface to generate a sense of touch of remote objects, we have developed a 1-DOF haptic interface, named Touch the Untouchable [2]. The system integrates a laser range finder (LRF) for measuring the distance from the interface to the remote objects, and a haptic apparatus, which has a haptic lever to expand or contract the virtual finger along the laser beam of the LRF. Although the user can feel the shape of the remote objects, the user cannot recognize the length of the virtual finger, which moves along with the haptic lever, correctly, because the Touch the Untouchable does not have a visual display. Then, we have developed a 1-DOF haptic interface [3], named Feelthrough. It has a haptic knob to change the length of the virtual finger infinitely and a see-through display on it to display the virtual finger to the user. The user can recognize the length of the virtual finger and feel the shape of remote objects more correctly. However, since the LRF and the haptic apparatus are coupled with each other and the LRF of the Feelthrough has 10-m measurement range, the distance to the touchable remote objects is limited within eye observation. Although the measurement accuracy is several centimeters, it is hard for the user to feel the details of the remote objects such as their textures.

In this study, to overcome this limitation, we propose a new hardware configuration and a haptic rendering algorithm of the Feelthrough. As a hardware configuration, the LRF and the haptic apparatus are separated. In comparison with the previous works, it is thought that the space constraint about a user's touching behavior can be reduced by using this method. At the remote site, a LRF is mounted on a 2-DOF pan and tilt camera platform for measuring the shape and texture of the remote objects. At the local site, a 1-DOF handheld haptic interface with a motion tracker is used for generating the feeling of touching the shape and texture of a remote object (Fig. 1). When the laser beam of the LRF and the virtual auxiliary line along the front direction of the interface intersect on a remote object, the distance between the interface and the object can be calculated. The position and orientation of the interface are not restricted to the LRF tightly. Therefore, the user can touch the remote object naturally as though he/she was touching the remote object from the normal direction of its surface, while the previous interfaces maintain the relative position and orientation of the LRF with respect to the interface. We developed a haptic rendering algorithm to realize natural touching behavior of a user.

3 Development of a Prototype System

The prototype system can be divided into two parts: the master-side interface and the slave-side device. The master-side interface consists of a 1-DOF handheld haptic interface for generating a reaction force in the depth direction to the ball of the thumb of a user, an electromagnetic motion tracking sensor, Isotrak II made by Polhemus, for measuring the motion of the interface, a stereoscopic liquid crystal display, FLATRON W2363D made by LG, for displaying 3D video images of the remote site and the CG image of the virtual finger, and a control PC (Fig. 2 left and center). As an end effector of the interface, a haptic knob coupled on a direct-drive DC motor maxon RE30 with an MR encoder is adopted. The slave device consists of a 2-DOF electric camera platform on which the LRF, LK-G500 made by Keyence, is mounted, a stereo camera that uses two CCD cameras, Scorpion made by PointGrey, and a control PC (Fig. 2 right). The platform is actuated by two digital servomotors, MX-28R made by Robotis Dynamixel. The slave system is connected to the master system via a network. Video images and LRF data are sent from the slave-side to the master-side. On the other hand, the control commands of the platform's posture, which are consistent with the position and orientation of the handheld interface, are sent from the master-side. In this prototype system, both sides are connected to the same PC directly in order to create an ideal environment without any communication delay.

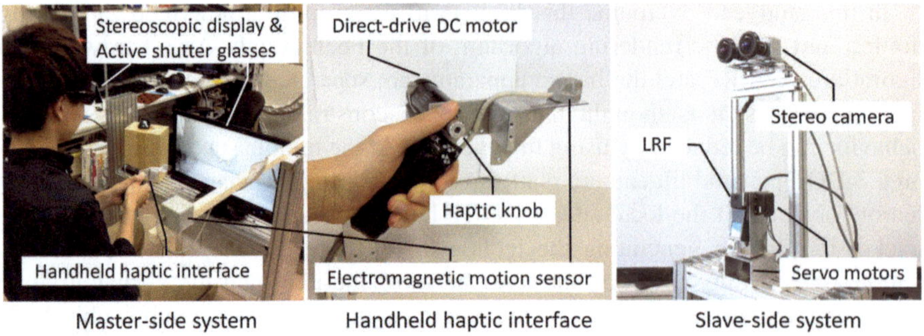

Fig. 2 Pictures of the prototype system

4 Haptic Rendering

The penalty method was applied to the haptic rendering algorithm. When the virtual fingertip invades the remote object, the haptic device generates a reaction force proportional to the amount of invasion (Fig. 3). Equation (1) shows the calculation method of the reaction force (T_{out}) where K_p means the proportional gain. Definitions of other values are shown in Fig. 3. In order to enable a user to operate a measurement point of the LRF with a natural hand motion, the platform is controlled to move the LRF measuring point to the intersection of the laser beam of the LRF and the extension of the virtual finger. The user can recognize a shape by perceiving a change in the reaction force at his/her fingertip according to a change in the distance with the movement of a measurement point on a remote object. Furthermore, the presentation of textures is possible in addition to shapes, by amplifying the high frequency component of the time series variation of the distance data and superimposing the data to the reaction force simultaneously.

$$T_{\text{out}} = \begin{cases} K_p(L_{\text{fin}} - L_{\text{obj}}) = K_p(L_{\text{fin}} - |\vec{V}_{\text{lrf}} - \vec{V}_{\text{hand}}|) & (L_{\text{fin}} - L_{\text{obj}} \geq 0) \\ 0 & (L_{\text{fin}} - L_{\text{obj}} < 0) \end{cases} \tag{1}$$

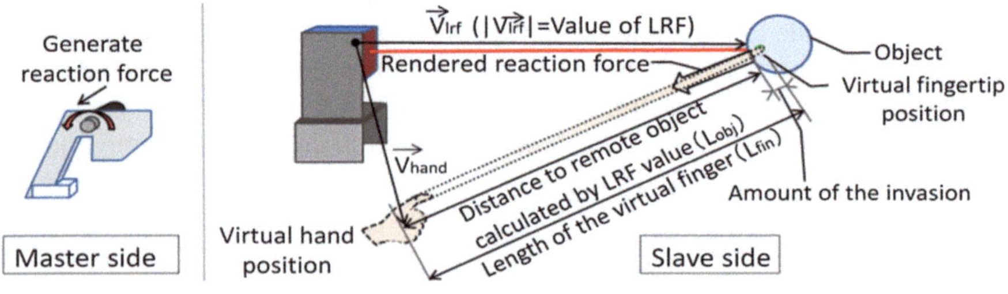

Fig. 3 Haptic rendering method

5 Evaluation

To evaluate the usefulness of the proposed method, a texture recognition test was conducted with 6 subjects. The subjects touched the textures on the side of two shapes, a cylinder and an edge-faced cube, with the prototype system (Fig. 4 left). The camera platform rotated the LRF according to the orientation of the handheld interface (this is the same as the previous method, named "w/o translation"). On the other hand, in the proposed method, the platform moved the laser beam of the LRF to the intersection along the direction of the front face of the interface (named "with translation"). 3 textures, which have different surface roughness, #60, #100, and #180 sandpapers, were used in this study. When the subjects felt one of the textures with each control method, they answered its estimated roughness they felt in comparison with the reference stimulus, which was #100 sandpaper, based on the magnitude estimation method. The subjects conducted 5 trials with each combination of shapes, textures, and control methods. Moreover, the positions of their virtual fingertip were recorded during each trial. Figure 4 right shows the coordinates of the touched objects.

The average perceptive roughness of 3 textures in the "with translation" are shown in Fig. 5. "**" means 1 % significance level ($p < 0.01$). Since significant differences among all textures are found, the subjects recognized the difference of 3 the textures. In addition, the histograms of the touch time on the coordinates are shown in Fig. 6. Each peak in the "with translation" condition is shifted toward the right as compared to the "w/o translation" condition. This means that the subjects

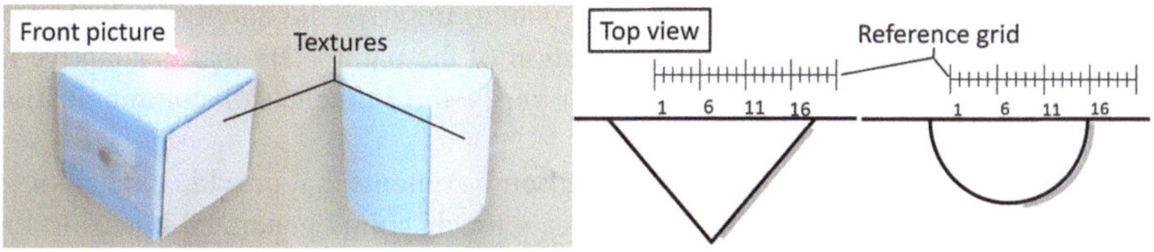

Fig. 4 Target objects

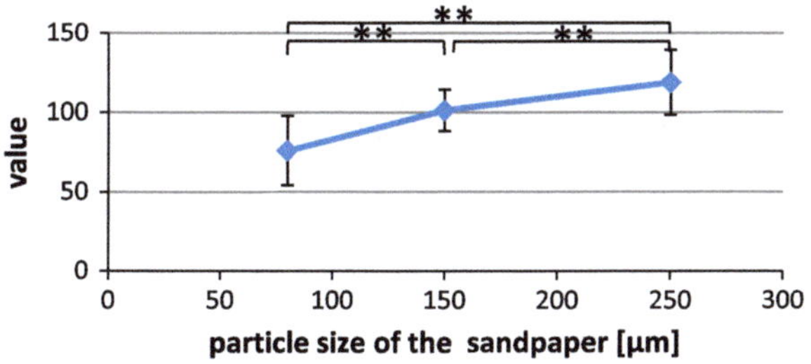

Fig. 5 Magnitude estimation of a perception of roughness on textures

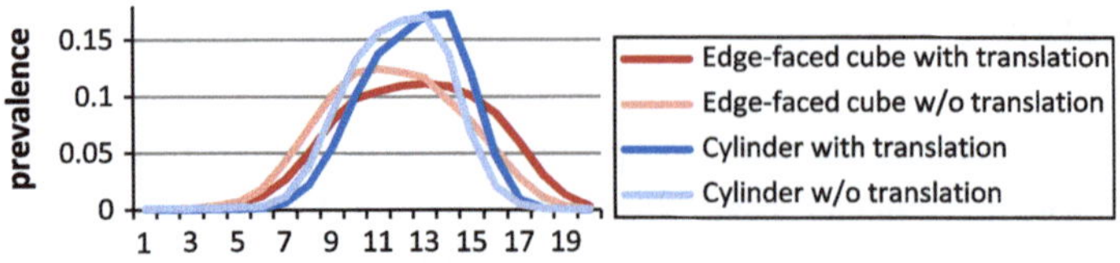

Fig. 6 Histogram of touch time in each method and for each shape

touched around the center of the texture of each shape in the "with translation" condition, while the subjects touched the position near to the center of each side surface of each object in the "w/o translation." Therefore, the proposed method enables users to touch the side of the remote objects easily and naturally.

6 Conclusion

This paper proposes a method to recognize the shape and texture of real objects at a remote site by using a low-DOF tele-existence system. By using the prototype system, we confirmed the following:

- A user can operate a virtual fingertip with his/her natural motion.
- He/she can recognize the three-dimensional surface shape of remote objects.
- He/she can recognize the difference in the texture of remote objects.

In this system, the difference in LRF-accuracy due to distance or color of the object is not considered. In the future, we plan to improve the quality of the haptic feedback by using an automatic determination of the parameters that is adaptive to these differences. In addition, we plan to verify the influence of communication delay on shape and texture perception.

Moreover, we aim to develop remote museum contents that provide explanations of the museum items according to the user's touch points generated by using the proposed system.

References

1. Fernando, C.L., Furukawa, M., Kurogi, T., Hirota, K., Kamuro, S., Sato, K., Minamizawa, K., Tachi, S.: TELESAR V: TELExistence surrogate anthropomorphic robot. In: SIGGRAPH 2012 Emerging Technologies, Article No. 23. Los Angeles, California (2012)
2. Yano, H., Miyamoto, Y., Iwata, H.: Touch the untouchable. In: SIGGRAPH ASIA 2009 Digital Experiences, p. 97 (2009)
3. Yano, H., Aoki, T., Iwata, H.: Handheld haptic interface for touching remote objects with visual display. In: Proceedings of Haptics 2012, pp. 349–354 (2012)

Generating Vibrotactile Images on the Human Palms

Keisuke Hasegawa and Hiroyuki Shinoda

Abstract We have created a system that generates vibrotactile images on the surface of bare human palms. It simultaneously projects visible images by a video projector and produces non-contact vibrotactile stimuli using focused airborne ultrasound on the users' skin. The vibrotactile images can be moved smoothly owing to the ultrasound phased array technique employed in the system. The vibrational texture can be tuned with amplitude modulation of ultrasound at the focal point. Our proposed system does not postulate that users wear specific devices for sensing tactile stimuli, which allows users to be free from any physical constraints.

Keywords Vibrotactile images · Airborne ultrasound · Multimodal display

1 Introduction

Among recent researches in providing multimodal experiences, the combination of visual and tactile stimuli has become one of the greatest issues in the context of psychophysical interests and its potential applications. Another significant progress in tactile displays is use of vibrotactile stimuli extracted from real-world vibration. Several researches have demonstrated that various and perhaps realistic tactile sensations can be aroused with those vibrations [1, 2].

Based on these advances, we propose a system that generates vibrotactile images on the surface of bare human palms using focused ultrasound superposed on the images by the video projector. Not only visible, but also tangible objects with variable textures are realized by the system.

K. Hasegawa (✉) · H. Shinoda
The University of Tokyo, 5-1-5 Kashiwanoha, 277-8561 Kashiwa-shi, Chiba-Ken, Japan
e-mail: Keisuke_Hasegawa@ipc.i.u-tokyo.ac.jp
URL: http://hapis.k.u-tokyo.ac.jp

H. Shinoda
e-mail: Hiroyuki_Shinoda@k.u-tokyo.ac.jp

H. Kajimoto et al. (eds.), *Haptic Interaction*, Lecture Notes in Electrical Engineering 277, DOI 10.1007/978-4-431-55690-9_56

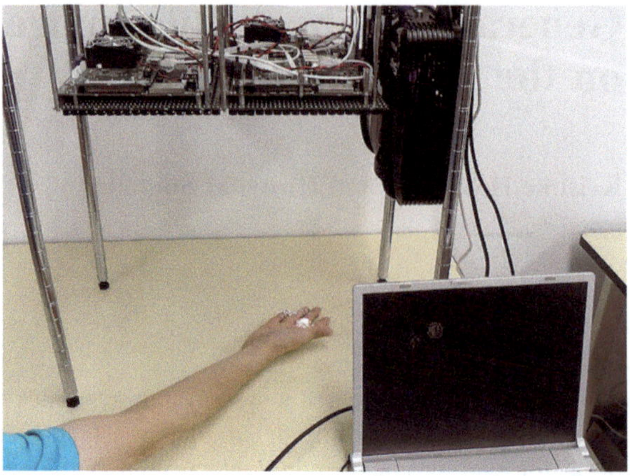

Fig. 1 System appearance

2 System Setup

Figure 1 shows the appearance of the proposed system. The system components are an ultrasound phased array, a video projector, and a PC for controlling the whole system. The ultrasound phased array is composed of planar arrays of ultrasound transducers connected to a controlling circuit [3, 4]. It can create a focus of ultrasound at an arbitrary location in the air. When an object, a palm in the proposed situation namely, is placed at the focal region, pressure is exerted [5]. The amplitude of the pressure can be time variant, which results in vibrational texture on the palm. By designing the temporal variance of focal amplitude, various vibrotactile textures can be presented to users [4]. Since the ultrasound is invisible, no visual interference is occurred when users feel tactile feelings. Owing to the property of ultrasound, a tangible image is naturally realized by simply projecting an image onto the ultrasound focus.

References

1. Yao, H.Y., Hayward, V.: An experiment on length perception with a virtual rolling stone. Proceedings of EuroHaptics, pp. 275–278 (2006)
2. Culbertson, H., Unwin, J., Goodman, B.E., Kuchenbecker, K.J.: Generating haptic texture models from unconstrained tool-surface interactions. IEEE World Haptics Conference, pp. 295–300 (2013)
3. Hoshi, T., Takahashi, M., Iwamoto, T., Shinoda, H.: Noncontact tactile display based on radiation pressure of airborne ultrasound. IEEE Trans. Haptics **3**(3), 155–165 (2010)
4. Hasegawa, K., Shinoda, H.: Aerial display of vibrotactile sensation with high spatial-temporal resolution using large-aperture airborne ultrasound phased array. In: Proceedings of IEEE World Haptics Conference (2013)
5. Awatani, J.: Studies on acoustic radiation pressure. I(General Considerations). J. Acoust. Soc. Am. **27**, 278–281 (1955)

A Proposal of Model-Based Haptization System for Animal Images

Takahiro Okubo, Katsuhito Akahane and Makoto Sato

Abstract In this paper, we propose a haptization system for animal images by using a 3D model of the animal. The haptization methods which were proposed before, there has been some problems. Some haptization methods are insufficient sense of presence, or require the depth information, so they are not fit to apply to animal images. Therefore, the purpose of this research is to improve the three-dimensional impression and the sense of presence by placing a 3D model of an animal on the image and performing the haptic feedback. But generating the 3D model of the animal from an image automatically and to estimate the pose is difficult. So we are going to make a system which have an existing 3D models and pose information, then users can choose and place them without stress.

Keywords Model-based haptization · Animal image · 3D model · Haptic device

1 Introduction

In recent years, the Internet has become a more familiar presence by the popularity of smartphones and tablet PCs. An opportunity to share images of the pet animals through social networking service or bulletin board system is increasing. When we saw these images, we can have various feelings, for example, the satisfaction of visual such as "it's cute," but it is not enough to meet the haptic interest such as "I want to touch."

T. Okubo (✉) · K. Akahane · M. Sato
Precision and Intelligence Laboratory, Tokyo Institute of Technology, Yokohama, Japan
e-mail: takahiro.okubo@hi.pi.titech.ac.jp

K. Akahane
e-mail: kakahane@hi.pi.titech.ac.jp

M. Sato
e-mail: msato@pi.titech.ac.jp

H. Kajimoto et al. (eds.), *Haptic Interaction*, Lecture Notes
in Electrical Engineering 277, DOI 10.1007/978-4-431-55690-9_57

So far, some methods of image haptization by using the haptic device have been proposed. Haptization means to make the haptic information presentation with respect to any data that cannot directly be touched by a human. However, these haptization methods are not suitable for systems for touching an animal in an image. The method of [1] can express the feeling of unevenness, but cannot represent the three-dimensional impression and sense of presence as if touching animals, because the presentable force is limited to reactive force of the surface of the image.

In this paper, we propose a haptization method using the 3D model of the animal to satisfy the haptic interest about animal images (we call it model-based haptization). And we implement a haptization system which is as if touching the animal being there. The system prepares the 3D model same as an animal of input image and places on the correct position.

2 Proposed System

Figure 1 shows a configuration of the proposed system. In the conventional system, it calculates the parameters for the force feedback from the information that the input image possesses, and then setting on the image, and presents the force when touching the image. In this system, the role of the input image is defined as the visual feedback, and the haptic feedback is performed from a 3D model of the animals. The system prepares a 3D model for haptic feedback in advance and sets the position and orientation in accordance with the animal in the image. At this time, it will not be rendered to the screen. Accordingly, it is possible to haptize two-dimensional animal image three dimensionally. As a side effect, it is also possible to touch the back of the animal which cannot be seen from the image, because it uses a 3D model.

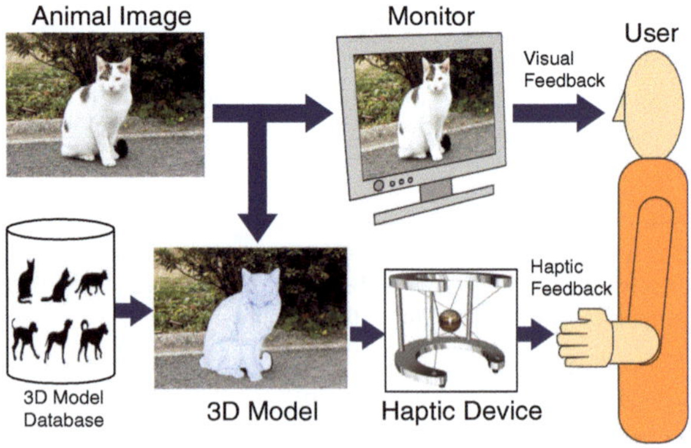

Fig. 1 System configuration

3 Experiment

In order to verify the effectiveness of the image haptization using a 3D model, we have carried out basic evaluation experiment both previous (image-based) haptization [1] and model-based haptization using Fig. 2.

Experimental results are shown in Fig. 3. The graph shows the average of scores that fifteen subjects answered. From experimental results, it can be seen that proposed method is better than color haptization method in all conditions. In particular, "sense of presence" was highly evaluated. This is because the object to touch is a

Fig. 2 An image used in the experiment

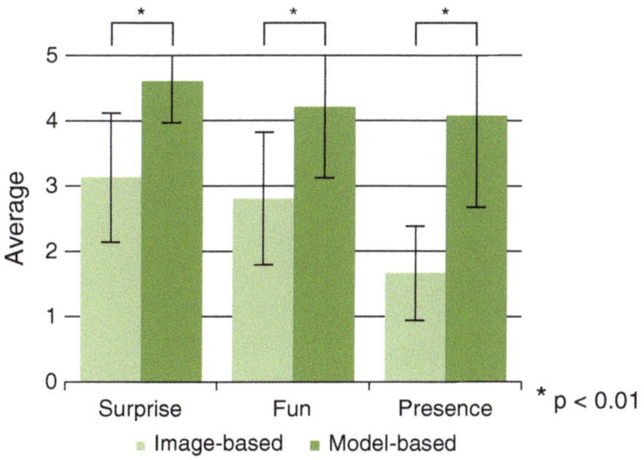

Fig. 3 Experimental result

3D model of the three-dimensional structure. Further, "surprise" and "fun" were also highly evaluated. It seems that the subjects were interested in touching two-dimensional image three dimensionally. In subjects' feedback, there were many requests of softness like a skin deformation.

Reference

1. Takahama, A., Akahane, K., Sato, M.: An image haptization method based on color and edge information. J. Inst. Image Electron. Eng. Jpn. **40**(4), 695–701 (2011)

Index

H. Kajimoto et al. (eds.), *Haptic Interaction*, Lecture Notes in Electrical Engineering 277, DOI 10.1007/978-4-431-55690-9

MIX
Papier aus verantwortungsvollen Quellen
Paper from responsible sources
FSC® C105338

If you have any concerns about our products,
you can contact us on
ProductSafety@springernature.com

In case Publisher is established outside the EU,
the EU authorized representative is:
Springer Nature Customer Service Center GmbH
Europaplatz 3, 69115 Heidelberg, Germany

Printed by Libri Plureos GmbH
in Hamburg, Germany